T0212645

SpringerBriefs in Mathematics

SpringerBriefs in Mathematics showcases expositions in all areas of mathematics and applied mathematics. Manuscripts presenting new results or a single new result in a classical field, new field, or an emerging topic, applications, or bridges between new results and already published works, are encouraged. The series is intended for mathematicians and applied mathematicians.

More information about this series at http://www.springer.com/series/10030

John C. George • Abdollah Khodkar • W.D. Wallis

Pancyclic and Bipancyclic Graphs

Springer

John C. George
Department of Mathematics
and Computer Science
Gordon State College
Barnesville, GA, USA

W.D. Wallis
Department of Mathematics
Southern Illinois University
Evansville, IN, USA

Abdollah Khodkar
Department of Mathematics
University of West Georgia
Carrollton, GA, USA

ISSN 2191-8198   ISSN 2191-8201   (electronic)
SpringerBriefs in Mathematics
ISBN 978-3-319-31950-6   ISBN 978-3-319-31951-3   (eBook)
DOI 10.1007/978-3-319-31951-3

Library of Congress Control Number: 2016935702

Printed on acid-free paper

This Springer imprint is published by Springer Nature
The registered company is Springer International Publishing AG Switzerland

The authors would like to dedicate this book to our families:
Amanda and Robert (JCG);
Sarah, Arvin, and Darian (AK);
Ann (WDW)

Preface

For nearly 50 years, there has been some interest in the cycles occurring as subgraphs of graphs. In 1971, Adrian Bondy introduced the idea of a *pancyclic* graph, one that contained cycles of every possible length. But from that time, the idea has been largely unexplored. Together with some of our colleagues, we have been looking at pancyclicity, and we decided it was time to write a book on the subject and related areas. It is our hope that students and researchers alike will find in this volume inspiration and ideas to facilitate their own work and shed further light on this fascinating topic.

We would like to express our thanks to some colleagues who have helped us in our endeavors, including Saad El-Zanati, Alison Marr, Alex Peterson, Nick Phillips, Christina Wahl, and Zach Walsh.

Our first draft was met with some very useful report from the referees, which we used to improve the book. We wish to thank them for their input.

We also wish to express our thanks to the staff of Springer and in particular to Razia Amzad for her help.

Detroit, MI, USA   John C. George
Carrollton, GA, USA   Abdollah Khodkar
Evansville, IN, USA   W.D. Wallis

Contents

1	**Graphs**	1
	1.1 Introduction	1
	1.2 Graphs: The Basics	1
	1.3 Products	3
	1.4 Walks, Paths, and Cycles	4
	1.5 Colorings and Cycles	6
2	**Degrees and Hamiltoneity**	9
	2.1 A Theorem of Chvátal	9
	2.2 A Theorem of Fan	10
	2.3 A Theorem of Bondy and Its Generalization	12
3	**Pancyclicity**	21
	3.1 Introduction	21
	3.2 Bounds	22
	3.3 Pancyclic Graph Products	32
	3.4 Open Problems	34
4	**Minimal Pancyclicity**	35
	4.1 Introduction	35
	4.2 Minimal Pancyclic Graphs: Small Orders	36
	4.2.1 Fewer Than Two Chords	37
	4.2.2 Two Chords	37
	4.2.3 Three Chords	38
	4.3 Four Chords	41
	4.4 Five Chords	42
	4.5 More General Bounds for Pancyclics	42
5	**Uniquely Pancyclic Graphs**	49
	5.1 Introduction	49
	5.2 Small Cases	49
	5.3 Outerplanar UPC Graphs	51
	5.4 More General UPC Graphs	53

	5.5	Cycle Space of a Graph	62
	5.6	Bounds on the Number of Edges in a UPC Graph	64
	5.7	Open Problems	67
6	**Bipancyclic Graphs**		**69**
	6.1	Introduction	69
	6.2	Edge Number Conditions	69
	6.3	Degree Conditions	71
7	**Uniquely Bipancyclic Graphs**		**81**
	7.1	Introduction	81
	7.2	Graphs with Fewer than Two Chords	82
	7.3	Two Chords	82
	7.4	Three Chords	84
	7.5	More Chords: Computer Searches	96
8	**Minimal Bipancyclicity**		**99**
	8.1	Introduction	99
	8.2	Minimal Bipancyclic Graphs with Excess Less than 2	100
	8.3	Excess 2	100
	8.4	Excess 3	101
	8.5	Excess 4	102
	8.6	More General Bounds for Bipancyclics	103
	8.7	Bipancyclic Graph Products	105
References			**107**

List of Figures

Fig. 1.1   Two representations of the graph K_4 3
Fig. 1.2   The graph $K_{3,5}$ 3

Fig. 2.1   Case where k is as large as possible................................ 11
Fig. 2.2   Case where $s \geq k + 2$ 11
Fig. 2.3   Case where $s = k + 1$ 12
Fig. 2.4 P with only one point (an endpoint) on C 13
Fig. 2.5 x' is the furthest neighbor of x................................... 14
Fig. 2.6 r and s are on the same path 14
Fig. 2.7 r and s are on different paths 14
Fig. 2.8   The C-path in case 1 15
Fig. 2.9   The C-path in case 2 15
Fig. 2.10   The required new cycle 15
Fig. 2.11   A C-path with a_1 and a_2 on C 16
Fig. 2.12   Case 1: $d_{C\setminus\{a\}}(x) = 0$....................................... 17
Fig. 2.13   Case 2: $d_{C\setminus\{a\}}(x) \geq 1$ and a^+ is not adjacent to $b^{+\ell}$ 17
Fig. 2.14   Case 2: $d_{C\setminus\{a\}}(x) \geq 1$ and a^+ is not adjacent to v^+ 18
Fig. 2.15 k parallel edges between H and C..................................... 18

Fig. 3.1   Graph F_n 31
Fig. 3.2   Constructing the $2k + 1$-cycle 33

Fig. 4.1   Cycles in the case of one chord 37
Fig. 4.2   Two chords 38
Fig. 4.3   The layout for nine vertices 38
Fig. 4.4   Cases of three chords 39
Fig. 4.5   Minimal pancyclic graphs up to order 14............................ 41
Fig. 4.6   Graph used to construct examples of orders 15–24 41
Fig. 4.7   Graph used to construct examples of orders 25–37 43

Fig. 5.1   Possibilities for two chords... 50
Fig. 5.2   UPC graphs with two or fewer chords 50

Fig. 5.3   The three non-isomorphic uniquely pancyclic graphs of
            order 14... 54
Fig. 5.4 C_1 does not skew to any chord; C_3 skews C_4 57
Fig. 5.5 C_1 skews C_3 only; C_3 does not skew C_4 57
Fig. 5.6 C_3 skews C_4; C_1 and C_4 are not adjacent 58
Fig. 5.7 C_3 skews C_4; C_1 and C_4 are adjacent 58
Fig. 5.8 C_1 skews to both C_3 and C_4 58

Fig. 6.1   Chords for Lemma 25.4 75
Fig. 6.2   *Dash lines* represent missing edges in the graph 79

Fig. 7.1   Cycles in the case of one chord 82
Fig. 7.2   The uniquely bipancyclic graph of order 8 82
Fig. 7.3   The possible cases with two chords................................. 83
Fig. 7.4   The possible cases with two chords................................. 84
Fig. 7.5   Cases of three chords 84
Fig. 7.6   Chord type AAAi 86
Fig. 7.7   Chord type AAAii 86
Fig. 7.8   Chord type AABi 87
Fig. 7.9   Chord type AABii 87
Fig. 7.10   Chord type AAC 89
Fig. 7.11   Uniquely bipancyclic graphs of type AAC: $d = 1,2,3,4$............ 90
Fig. 7.12   Chord type ABBi 91
Fig. 7.13   Chord type ABBii 91
Fig. 7.14   Chord type ABC 91
Fig. 7.15   Uniquely bipancyclic graphs of type ABC 92
Fig. 7.16   Chord type ACC 93
Fig. 7.17   Chord type BBBii... 94
Fig. 7.18   Chord type BBC 94
Fig. 7.19   Chord type BCC 94
Fig. 7.20   The uniquely bipancyclic graphs of order less than 32 95
Fig. 7.21   Layout for the six non-isomorphic UBPC graphs of
            order 44... 96

Fig. 8.1   Minimal bipancyclic graphs up to order 8 100
Fig. 8.2   Minimal bipancyclic graphs of orders 10–14........................ 100
Fig. 8.3   Minimal bipancyclic graphs of orders 16–26........................ 101
Fig. 8.4   Chord patterns ACC and CCC 101
Fig. 8.5   A model for 28–44 vertices 102

Chapter 1
Graphs

1.1 Introduction

In this chapter we shall survey some basic ideas of graph theory. Most of our readers will already be familiar with these topics; those who are not, or wish to explore them further or to see proofs, should consult a recent book on the subject, such as [34] or [36]. Our main aim here is to ensure that we use all the terminology in the same way, and to standardize our notation.

1.2 Graphs: The Basics

A *graph G* consists of a finite set $V(G)$ of objects called *vertices* together with a set $E(G)$ of unordered pairs of vertices; the elements of $E(G)$ are called *edges*. We write $v = v(G)$ and $e = e(G)$ for the orders of $V(G)$ and $E(G)$, respectively; these are called the *order* and *size* of G. In terms of the more general definitions sometimes used, we can say that "our graphs are finite and contain neither loops nor multiple edges." A *multigraph* is defined in the same way as a graph except that there may be more than one edge corresponding to the same unordered pair of vertices. The edge containing x and y is written (x, y) or xy; we refer to x and y as the *endpoints* of (x, y), and we say this edge *joins* x to y. $G - (x, y)$ denotes the result of deleting edge (x, y) from G; if x and y were not adjacent, then $G + (x, y)$ is the graph constructed from G by adding an edge (x, y). Similarly $G - x$ is the graph derived from G by deleting one vertex x (and all the edges on which x lies), and $G - S$ denotes the result of deleting some set S of vertices.

If vertices x and y are endpoints of one edge in a graph, then x and y are said to be *adjacent* to each other, and it is often convenient to write $x \sim y$. The set of all vertices adjacent to x is called the *neighborhood* of x, and denoted $N(x)$. We define the *degree* or *valency* $d(x)$ of the vertex x to be the number of edges that have x as an

© The Author(s) 2016
J.C. George et al., *Pancyclic and Bipancyclic Graphs*, SpringerBriefs
in Mathematics, DOI 10.1007/978-3-319-31951-3_1

endpoint; if more than one graph is involved, we denote the degree of x in graph G by $d_G(x)$. If $d(x) = 0$, x is an *isolated* vertex, and a vertex x with $d(x) = 1$ is a *leaf* or *pendant vertex*. A graph is called *regular* if all its vertices have the same degree. We shall write $\delta(G)$ for the smallest of all degrees of vertices of G, and $\Delta(G)$ for the largest. (We also write $\Delta(G)$ for the common degree of a regular graph G.) If G has v vertices, so that its vertex set is, say,

$$V(G) = \{a_1, a_2, \ldots, a_v\},$$

then its *adjacency matrix* M_G is the $v \times v$ matrix with entries m_{ij}, such that

$$m_{ij} = \begin{cases} 1 & \text{if } a_i \sim a_j, \\ 0 & \text{otherwise.} \end{cases}$$

A vertex and an edge are called *incident* if the vertex is an endpoint of the edge, and two edges are called incident if they have a common endpoint. A set of edges is called *independent* if no two of its members are incident, while a set of vertices is independent if no two of its members are adjacent.

Theorem 1. *In any graph or multigraph, the number of edges equals half the sum of the degrees of the vertices.*

Corollary 1.1. *In any graph or multigraph, the number of vertices of odd degree is even. In particular, a regular graph of odd degree has an even number of vertices.*

If G is a graph, it is possible to choose some of the vertices and some of the edges of G in such a way that these vertices and edges again form a graph, H say. H is then called a *subgraph* of G, and is denoted $H \leq G$. Clearly every graph G has itself and the 1-vertex graph (which we shall denote K_1) as subgraphs; we say H is a *proper* subgraph of G if it neither equals G nor K_1. If U is any subset of vertices of G, then the subgraph consisting of U and all the edges of G that join two vertices of U is called an *induced* subgraph, the *subgraph induced by U*, and is denoted $\langle U \rangle$. A subgraph G of a graph H is called a *spanning* subgraph if $V(G) = V(H)$.

Given a set S of v vertices, the graph formed by joining all pairs of members of S is called the *complete* graph on S, and denoted K_S. We also write K_v to mean any complete graph with v vertices. The set of all edges of $K_{V(G)}$ that are *not* in a graph G will form a graph with $V(G)$ as vertex set; this new graph is called the *complement* of G, and is denoted $\overline{G}$. More generally, if G is a subgraph of H, then the graph formed by deleting all edges of G from H is called the *complement of G in H*, denoted $H - G$.

To represent a graph in a diagram, use dots for the vertices, and a line joining two dots for an edge with the two corresponding vertices as endpoints. The lines are usually, but not necessarily, straight. Figure 1.1 shows two diagrams representing the same graph.

A graph is called *planar* if it can be drawn in such a way that no two edges cross. Figure 1.1 shows that K_4 is planar. It also shows that sometimes a planar graph can

Fig. 1.1 Two representations of the graph K_4

be represented by a drawing in which some edges cross; but if the graph is planar there must always be at least one drawing in which no crossings occur.

An *isomorphism* of a graph G onto a graph H is a one-to-one map ϕ from $V(G)$ onto $V(H)$ with the property that a and b are adjacent vertices in G if and only if $\phi(a)$ and $\phi(b)$ are adjacent vertices in H; G is isomorphic to H if and only if there is an isomorphism of G onto H. From this definition it follows that all complete graphs on n vertices are isomorphic; the notation K_n can be interpreted as being a generic name for the typical representative of the isomorphism-class of all n-vertex complete graphs.

A graph is *disconnected* if its vertex set can be partitioned into two subsets, V_1 and V_2, which have no common element, in such a way that there is no edge with one endpoint in V_1 and the other in V_2; if a graph is not disconnected, then it is *connected*. A disconnected graph consists of a number of disjoint subgraphs; a maximal connected subgraph is called a *component*.

A *bipartite graph* is a graph with two disjoint sets of vertices, V_1 and V_2 say, where no two vertices in the same set are adjacent. The two sets V_1 and V_2 are called the *components* of the bipartite graph. An important case is the *complete bipartite graph* based on V_1 and V_2, in which all vertices in V_1 are adjacent to all vertices in V_2. If there are m vertices in one set and n in the other, the complete bipartite graph is denoted $K_{m,n}$. Figure 1.2, below, shows an example of the bipartite graph $K_{3,5}$. A bipartite graph is called *balanced* if the two sets V_1 and V_2 are equal in size.

Fig. 1.2 The graph $K_{3,5}$

1.3   Products

Graph products of graphs G and H are usually defined on the vertex set $V(G) \times V(H)$, and have edges between vertices $g_1 h_1$ and $g_2 h_2$ determined by conditions on the adjacency of g_1 and g_2 in G, and h_1 and h_2 in H. For an extensive discussion of graph products, the reader is referred to [18]. For our purposes, we need to define the *cartesian product*, the *tensor product*, and the *strong product*.

The *cartesian product*, denoted $G \times H$, has an edge between $g_1 h_1$ and $g_2 h_2$ when either ($g_1 = g_2$ and $h_1 \sim h_2$) or ($h_1 = h_2$ and $g_1 \sim g_2$). Of particular interest is the construction $G \times K_2$, called a *prism*. Note that the product $G \times H$ is bipartite if G and H are bipartite.

The *tensor product*, also called the *Kronecker product* or *conjunction*, is the product $G \otimes H$ with an edge between $g_1 h_1$ and $g_2 h_2$ when ($g_1 \sim g_2$ and $h_1 \sim h_2$). The product possesses two properties of interest here. First, if either G or H is bipartite, then $G \otimes H$ is also bipartite. Second, if both are bipartite, then the product is disconnected. (If G and H are connected and bipartite, then $G \otimes H$ has exactly two components.)

The *strong product* $G \boxtimes H$ is the graph whose edge set is the union of the edge sets of $G \times H$ and $G \otimes H$. We denote by $G^{\boxtimes m}$ the strong product of m copies of G.

1.4 Walks, Paths, and Cycles

A *walk* in a graph G is a finite sequence of vertices $a_0, a_1, \ldots, a_n$ and edges $e_1, e_2, \ldots, e_n$ of G:

$$a_0, e_1, a_1, e_2, a_2, \ldots, e_n, a_n,$$

where the endpoints of e_i are a_{i-1} and a_i for each i. A *simple walk* is a walk in which no edge is repeated. A *path* is a walk in which no vertex is repeated; the *length* of a path is its number of edges. It is not hard to see that if a graph contains a walk from a_0 to a_n then it contains a path from a_0 to a_n: if the walk contains a duplicated vertex, say $a_i = a_j, j > i$, simply omit $a_i, e_{i+1}, a_{i+1}, e_{i+2}, \ldots, e_j$; repeat until there are no duplicates.

A walk is *closed* when the first and last vertices, a_0 and a_n, are equal. A *cycle* of length n is a closed simple walk of length n, $n \geq 3$, in which the vertices $a_0, a_1, \ldots, a_{n-1}$ are all different. A convenient notation for a walk with vertices (in order) $a_0, a_1, \ldots, a_n$ is $(a_0, a_1, \ldots, a_n)$, while the corresponding cycle is denoted $(a_0, a_1, \ldots, a_n, a_0)$.

When x and y are vertices in the same component of a graph, we define the *distance* $D(x, y)$ between x and y to be the length of the shortest path from x to y.

Cycles give the following useful characterization of bipartite graphs:

Theorem 2. *A graph is bipartite if and only if it contains no cycle of odd length.*

Proof. (i) Suppose G is a bipartite graph with disjoint vertex sets V_1 and V_2. Suppose G contains a cycle of odd length, with $2k + 1$ vertices

$$a_1, a_2, a_3, \ldots, a_{2k+1},$$

where a_i is adjacent to a_{i+1} for $i = 1, 2, \ldots, 2k$, and a_{2k+1} is adjacent to a_1. Suppose a_i belongs to V_1. Then a_{i+1} must be in V_2, as otherwise we would have two adjacent vertices in V_1; and conversely. So, if we assume $a_1 \in V_1$ we get, successively, $a_2 \in V_2, a_3 \in V_1, \ldots, x_{2k+1} \in V_1$. Now $a_{2k+1} \in V_1$ implies $a_1 \in V_2$, which contradicts the disjointness of V_1 and V_2.

(ii) Suppose that G is a graph with no cycle of odd length. Without loss of generality we need only consider the case where G is connected. Choose an arbitrary vertex x in G, and partition the vertex set by defining V_1 to be the set of vertices whose distance from x is even, and V_2 to be the set of vertices whose distance from x is odd; x itself belongs to V_1.

Now select two vertices b_1 and b_2 in V_1. Let P be a shortest path from x to b_1 and Q a shortest path from x to b_2. Denote by u the last vertex common to P and Q. Since P and Q are shortest paths, so are their sections from x to u, which therefore have the same length. Since the lengths of both P and Q are even, the lengths of their sections from u to b_1 and u to b_2, respectively, have equal parity, so the path from b_1 to u (in reverse direction along P) to b_2 (along Q) has even length.

If $b_1 \sim b_2$, then this path together with the edge $b_1 b_2$ gives a cycle of odd length, which is a contradiction. Hence no two vertices in V_1 are adjacent. Similarly no two vertices in V_2 are adjacent, and G is a bipartite graph.   □

A graph that contains no cycles at all is called *acyclic*; a connected acyclic graph is called a *tree*. Clearly all trees are bipartite graphs. Vertices of a tree other than leaves are called *internal*. One important example of a tree is the *n-star* $K_{1,n}$; the vertex of degree n is called the *center* of the star. A union of disjoint trees is called a *forest*.

Obviously the set of vertices and edges that constitute a path in a graph is itself a graph, and the word "path" is used for this graph: we define the *path* P_n to be a graph with n vertices $a_1, a_2, \ldots, a_n$ and $n - 1$ edges $a_1 a_2, a_2 a_3, \ldots, a_{n-1} a_n$. A *cycle* C_n is defined similarly, except that the edge $a_n a_1$ is also included, and (to avoid the triviality of allowing K_2 to be defined as a cycle) n must be at least 3. The latter convention ensures that every C_n has n edges.

A path that passes through every vertex in a graph is called a *Hamilton path* and a cycle that passes through every vertex is called a *Hamilton cycle*. A graph with a Hamilton cycle is called *Hamiltonian*. A graph with no such cycle is called *non-Hamiltonian*, and a non-Hamiltonian graph is *maximally non-Hamiltonian* if the addition of any edge will result in a Hamiltonian graph. If enough edges are added to a graph with at least 3 vertices, the resulting graph will be Hamiltonian, because eventually the complete graph, which is Hamiltonian, will be obtained. So obviously every non-Hamiltonian graph is a subgraph of a maximally non-Hamiltonian graph.

Suppose a bipartite graph with components V_1 and V_2 is Hamiltonian. Since no two vertices of the same component are adjacent, every second vertex in the Hamilton cycle must be in the same component. As all vertices are included, this shows that a Hamiltonian bipartite graph must be balanced.

Finally, suppose a graph contains cycles of all possible length. For example, if the graph has n vertices, it contains a cycle of length k for every integer k, $3 \le k \le n$.

Such a graph is called *pancyclic*. As we said in the Preface, our main interest in this book is pancyclic graphs and related ideas. We shall return to this topic in Chap. 3, after some more preliminaries.

1.5 Colorings and Cycles

A *coloring* (or *vertex-coloring*) of a graph is a way of applying a set of labels to the vertices. The labels are called *colors* because one convenient way of representing such a labeling is by coloring the vertices in a diagram. A coloring is called *proper* if no two adjacent vertices receive the same color. The result of coloring is to divide the set of vertices into disjoint subsets, where each subset consists of the vertices that receive a specific color. These subsets are called *color classes*. A *proper coloring* of G is a coloring in which no two adjacent vertices belong to the same color class. In symbols, if $\xi(x)$ denotes the color assigned to vertex x, then

$$x \sim y \Rightarrow \xi(x) \neq \xi(y).$$

A proper coloring is called an *r-coloring* if it uses r colors; if G has an r-coloring, then G is called *r-colorable*. The *chromatic number* $\chi(G)$ of a graph G is the smallest integer r such that G has an r-coloring. A coloring of G in $\chi(G)$ colors is called *minimal*. The phrase "G is *r-chromatic*" means that $\chi(G) = r$.

A cycle of length v has chromatic number 2 if v is even and 3 if v is odd. The graph $K_{1,n}$ has chromatic number 2. Clearly $\chi(G) = 1$ if G has no edges and $\chi(G) \geq 2$ if G has at least one edge. The complete graph on n vertices has chromatic number n.

There are two useful results when looking for chromatic numbers. The first is that, if the graph G has a subgraph H, then $\chi(G)$ cannot be smaller than $\chi(H)$. The second, which we shall use in Chap. 3, is Brooks' Theorem:

Theorem 3 ([6]). *If G is connected, then $\chi(G) \leq \Delta(G)$ unless G is a complete graph or an odd cycle.*

Proof. We shall assume G is neither a complete graph nor an odd cycle. We shall write Δ for $\Delta(G)$. If $\Delta \leq 2$, then the only possibility is a path or even cycle, and the result is easy, so we assume $\Delta \geq 3$. Our proof proceeds by induction on Δ, and, for each Δ, we will use induction on n. The induction starts at $n = \Delta + 1$, and the theorem is true in this case, since if $|G| = n = \Delta + 1$ and $G \neq K_n$, we can color G with Δ colors by using the same color for some two non-adjacent vertices. Therefore, suppose $n \geq \Delta + 2$. We consider three cases.

Case 1: There is a vertex v such that $G - v$ is disconnected. Let the components of $G - v$ be $G_1, G_2, \ldots, G_t$. Consider the subgraphs of G induced by the vertex sets $V(G_1) \cup \{v\}, V(G_2) \cup \{v\}, \ldots, V(G_t) \cup \{v\}$. We may Δ-color each of these graphs by induction (if one of the graphs is complete or an odd cycle, its maximum degree must be strictly less than Δ). Switching colors within some of these colorings if necessary, we may assume that v receives color 1 in all t colorings, which we can therefore combine to get a Δ-coloring of G.

Case 2: $G - v$ is connected for all v, but there are two non-adjacent vertices v and w such that $G - \{v, w\}$ is disconnected. Let A be a component of $G - \{v, w\}$, and let $B = V(G) \setminus (V(A) \cup \{v, w\})$. If there are no edges from v to A, then $G - w$ is disconnected, which we are assuming is not the case. Therefore, there is at least one edge from v to A. Similarly, there is at least one edge from w to A, at least one edge from v to B, and at least one edge from w to B. Let G_1 be the subgraph of G induced by vertices $V(G) \setminus B$ and let G_2 be the subgraph of G induced by the vertices $V(G) \setminus V(A)$. It is tempting at this point to Δ-color G_1 and G_2 by induction and then combine the colorings, but it may not be possible to combine the colorings (to see why, consider the case when G is an odd cycle).

Instead, we note that, from the above observations, v and w have degree at most $\Delta - 1$ in both G_1 and G_2, so that we may Δ-color $G_3 = G_1 + vw$ and $G_4 = G_2 + vw$ by induction, unless one of them is complete (if either of them is an odd cycle, we can Δ-color it since $\Delta > 2$). Such colorings, if they exist, can be combined because v and w will be forced to have different colors in both of them. We can then switch colors if necessary to ensure that v and w are colored 1 and 2, respectively, in both colorings.

If G_3 is a clique on $\Delta + 1$ vertices, then each of v and w must have degree 1 in G_2 (since both have degree Δ in G_3 and $\Delta - 1$ in G_1). In G_2, we can combine v and w into a single vertex, obtaining a graph G_5, which can be Δ-colored by induction. Therefore, there are Δ-colorings of both G_1 and G_2 in which both v and w get the same color. These colorings can be combined to provide a Δ-coloring of G.

Case 3: $G - \{v, w\}$ is connected for every pair of non-adjacent vertices v and w. Select a vertex u of maximum degree Δ. Since $G \neq K_n$, some pair of neighbors v and w of u are not adjacent. We define $v_1 = v$, $v_2 = w$, $v_n = u$ and, working backwards from v_{n-1} to v_3, we ensure that each v_i has some neighbor among $\{v_{i+1}, v_{i+2}, \ldots, v_n\}$, this is possible since $G - \{v, w\}$ is connected. Running the greedy algorithm with this ordering of the vertices, we see that $v_1 = v$ and $v_2 = w$ both get color 1, and also that we never need to use color $\Delta + 1$ on $v_3, v_4, \ldots, v_{n-1}$, since each such v_i has only at most $\Delta - 1$ neighbors among the already colored vertices. Finally, when we come to color v_n, two of its Δ neighbors have received color 1, so that one of the colors $1, 2 \ldots, \Delta$ is available to color v_n itself. This completes the induction step. $\qquad\square$

Chapter 2
Degrees and Hamiltoneity

Obviously any pancyclic graph must contain a Hamilton cycle. In this chapter we look at some results relating the Hamiltonian property to the set of degrees of vertices in the graph.

2.1 A Theorem of Chvátal

In 1972, Chvátal [8] proved the following theorem :

Theorem 4. *Suppose G is a graph with vertices* $\{a_1, a_2, \ldots, a_v\}$, $v \geq 3$, *whose degrees satisfy*

$$d(a_1) \leq d(a_2) \leq \ldots, \leq d(a_v).$$

If

$$d(a_k) \leq k < \tfrac{1}{2}v \Rightarrow d(a_{v-k}) \geq v - k, \tag{2.1}$$

then G is Hamiltonian.

Proof. Suppose there is a graph G that satisfies (2.1) but is not Hamiltonian. Then G will be a subgraph of a maximally non-Hamiltonian graph, G^* say, and G^* will also satisfy (2.1).

Suppose p and q are non-adjacent vertices in G^* such that $d(p) + d(q)$ is as large as possible; for convenience, say $d(p) \leq d(q)$. Note that this implies that if r is another vertex not adjacent to q, then $d(r) \leq d(p)$.

If the edge (q, p) is added to G^*, then the resulting graph will be Hamiltonian, and every Hamilton cycle must contain (q, p) (otherwise the cycle would be in G^*); say the cycle is $(a_1, a_2, \ldots, a_v, a_1)$, where $a_1 = p$ and $a_v = q$. Define two sets S and T:

© The Author(s) 2016
J.C. George et al., *Pancyclic and Bipancyclic Graphs*, SpringerBriefs
in Mathematics, DOI 10.1007/978-3-319-31951-3_2

$$S = \{i : (p, a_{i+1}) \text{ is an edge in } G^*\}, \quad T = \{i : (a_i, q) \text{ is an edge in } G^*\}.$$

Sets S and T must be disjoint: if j were a common element, then

$$(a_j, a_{j-1}, \ldots, a_1, a_{j+1}, a_{j+2}, \ldots, a_v)$$

would be a Hamilton cycle in G^*. So $S \cap T = \emptyset$, whence $|S| + |T| = |S \cup T| \leq v - 1$. Now $d(p) = |S|$ and $d(q) = |T|$, so $d(p) + d(q) = |S| + |T| < v$ and $d(p) < \frac{1}{2}v$. Moreover, as $S \cap T = \emptyset$, no a_j with $j \in S$ is adjacent to q, so $d(a_j) \leq d(p)$ for all members of S. So there are at least $d(p)$ vertices whose degrees do not exceed $d(p)$. Setting $k = d(p)$ we have $d(x_k) \leq k < \frac{1}{2}v$, so from (2.1), $d(x_{v-k}) \geq v - k$. So there are at least $k + 1$ vertices with degree at least $v - k$. As p has degree k, it cannot be adjacent to all of them. So there is a vertex, r say, with $d(r) \geq v - k$, which is not adjacent to p. But then $d(p) + d(r) \geq v > d(p) + d(q)$, contradicting the maximality of $d(p) + d(q)$. So the theorem is true by contradiction. □

If G is any graph, then a *factor* or *spanning subgraph* of G is a subgraph with vertex set $V(G)$. A *factorization* of G is a set of factors of G that are pairwise *edge-disjoint*—no two have a common edge—and whose union is all of G.

Every graph has a factorization, quite trivially: since G is a factor of itself, $\{G\}$ is a factorization of G. However, it is more interesting to consider factorizations in which the factors satisfy certain conditions. In particular a *one-factor* is a factor that is a regular graph of degree 1. In other words, a one-factor is a set of pairwise disjoint edges of G that between them contain every vertex. A *one-factorization* of G is a decomposition of the edge set of G into edge-disjoint one-factors. Similarly a *two-factor* is a factor that is a regular graph of degree 2—a union of disjoint cycles—and a *two-factorization* of G is a decomposition of the edge set of G into edge-disjoint two-factors.

2.2   A Theorem of Fan

The following theorem was proven by Geng-Hua Fan [11] in 1984:

Theorem 5. *Let G be a 2-connected graph on $n \geq 3$ vertices. If $D(x, y) = 2$ implies* $\max\{d(x), d(y)\} \geq n/2$ *for all vertices x and y, then G is Hamiltonian.*

Proof. The proof is by contradiction. Let G be a graph satisfying the given condition and G has no Hamilton cycle. Let $P = (v_0, v_1, \ldots, v_m)$ be a longest path in G of length m, chosen such that $d(v_0) + d(v_m)$ is as large as possible. If $d(v_0) + d(v_m) \geq n$, then there are at least two consecutive vertices on P, say v_i and v_{i+1}, such that $v_i v_m, v_{i+1} v_0 \in E(G)$. This gives us a cycle of length $m + 1$. Since G is 2-connected, we have either a Hamilton cycle or a path of length $m + 1$. Both lead to contradictions. Hence, $d(v_0) + d(v_m) < n$. Without loss of generality, we may assume $d(v_0) < n/2$.

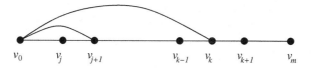

Fig. 2.1 Case where k is as large as possible

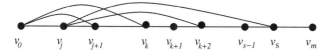

Fig. 2.2 Case where $s \geq k + 2$

An argument similar to that described above shows that:

Claim 1. G has no cycle of length $m + 1$.

Since G is 2-connected, v_0 is joined to at least one vertex on P other than v_1, say v_k. We choose k as large as possible (see Fig. 2.1). Clearly, $k \geq 2$ and $k < m$, otherwise we have a cycle of length $m + 1$, which contradicts Claim 1.

Claim 2. $v_0 v_i \in E(G)$ for all $1 \leq i \leq k$.

Proof. As $v_0 v_k \in E(G)$, there is another longest path $(v_{k-1}, v_{k-2}, \ldots, v_0, v_k, \ldots, v_m)$ in G. By the maximality of $d(v_0) + d(v_m)$, $d(v_{k-1}) < d(v_0) < n/2$. So $\max\{d(v_{k-1}), d(v_0)\} < n/2$. Hence, $d(v_0, v_{k-1}) \neq 2$. However, (v_0, v_k, v_{k-1}) is a path of length 2 and thus we must have $v_0 v_{k-1} \in E(G)$. If $k - 1 = 1$ we stop, if not we repeat this process and thus we obtain that $v_0 v_i \in E(G)$ for all $1 \leq i \leq k$.

Claim 3. $d(v_i) \leq d(v_0)$ for all $1 \leq i \leq k - 1$.

Proof. Suppose that $d(v_j) > d(v_0)$ for some $1 \leq j \leq k-1$. Since $v_0 v_{j+1} \in E(G)$, by Claim 2, $(v_j, v_{j-1}, \ldots, v_0, v_{j+1}, \ldots, v_k, \ldots, v_m)$ is another longest path with $d(v_j) + d(v_m) > d(v_0) + d(v_m)$, a contradiction.

Claim 4. $d(v_{k+1}) > d(v_0)$.

Proof. Since $v_0 v_{k+1} \in E(G)$ by the choice of v_k and the path (v_0, v_k, v_{k+1}) is of length 2, it follows that $d(v_0, v_{k+1}) = 2$. Hence, $\max\{d(v_0), d(v_{k+1})\} \geq n/2$. So $d(v_{k+1}) \geq n/2 > d(v_0)$.

For every $1 \leq i \leq k-1$, the vertex v_i cannot be joined to any vertex outside P by Claim 2. Since G is 2-connected, it follows that there exists an edge $v_j v_s$ in G such that $j < k < s$. We choose $v_j v_s$ in G such that $j < k < s$ and s is as large as possible. We now consider two cases.

Case 1. $s \geq k + 2$ (see Fig. 2.2). By Claim 2, $v_0 v_{j+1} \in E(G)$ and so there is a longest path $(v_{s-1}, v_{s-2}, \ldots, v_{j+1}, v_0, \ldots, v_j, v_s, \ldots, v_m)$ in G. It follows that $d(v_{s-1}) \leq d(v_0) < n/2$ by the maximality of $d(v_0) + d(v_m)$. On the other hand, $d(v_j) \leq d(v_0) < n/2$ by Claim 3. Therefore $\max\{d(v_{s-1}), d(v_j)\} < n/2$. Hence, $d(v_j, v_{s-1}) \neq 2$. But (v_j, v_s, v_{s-1}) is a path of length 2 and thus we

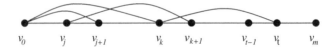

Fig. 2.3 Case where $s = k + 1$

must have that $v_j v_{s-1} \in E(G)$. If $s - 1 > j + 1$, we have another longest path $(v_{s-2}, v_{s-3}, \ldots, v_{j+1}, v_0, \ldots, v_j, v_{s-1}, v_s, \ldots, v_m)$ in G. Repeat this process to obtain $v_j v_{s-2} \in E(G)$. If $s - 2 > j + 1$ we go through the same process again. Consequently, we have $v_j v_i \in E(G)$ for all $j + 1 < i < s$. In particular, $v_j v_{k+2} \in E(G)$ because $s \geq k + 2$. This means that there is a longest path

$$(v_{k+1}, v_k, \ldots, v_{j+1}, v_0, \ldots, v_j, v_{k+2}, \ldots, v_m)$$

in G with $d(v_{k+1}) + d(v_m) > d(v_0) + d(v_m)$, a contradiction.

Case 2. $s = k + 1$ (see Fig. 2.3). Note that $(v_k, v_{k-1}, \ldots, v_{j+1}, v_0, \ldots, v_j, v_{k+1}, \ldots, v_m)$ is a longest path and so by the maximality of $d(v_0) + d(v_m)$ we obtain $d(v_k) \leq d(v_0) < n/2$.

If $k + 1 = m$, we obtain a cycle $(v_0, v_1, \ldots, v_j, v_{k+1}, v_k, \ldots, v_{j+1}, v_0)$ of length $m + 1$. This contradicts Claim 1. Now assume $k + 1 < m$. It follows from the 2-connectedness of G and the choice of v_s that there must be $v_k v_t \in E(G)$ such that $t \geq k + 2$. This implies that there is another longest path $(v_{t-1}, v_{t-2}, \ldots, v_{k+1}, v_j, \ldots, v_0, v_{j+1}, \ldots, v_k, v_t, \ldots, v_m)$ in G. By the maximality of $d(v_0) + d(v_m)$ we must have $d(v_{t-1}) \leq d(v_0) < n/2$. Hence, $\max\{d(v_k), d(v_{t-1})\} < n/2$. Therefore $d(v_k, v_{t-1}) \neq 2$. Now note that (v_k, v_t, v_{t-1}) is a path of length 2. So we obtain $v_k v_{t-1} \in E(G)$. If $t - 1 > k + 1$ we repeat this process to obtain $v_k v_t \in E(G)$ for all $k + 1 \leq i \leq t$. In particular, $v_k v_{k+2} \in E(G)$ because $t \geq k + 2$. Hence, there is a longest path $(v_{k+1}, v_j, \ldots, v_0, v_{j+1}, \ldots, v_k, v_{k+2}, \ldots, v_t, \ldots, v_m)$ in G. By Claim 4, $d(v_{k+1}) + d(v_m) > d(v_0) + d(v_m)$, a contradiction. This completes the proof. $\square$

2.3   A Theorem of Bondy and Its Generalization

The following result is due to Bondy [5]:

Theorem 6. *Let G be a 2-connected graph on n vertices. Suppose that for every set of three independent vertices x, y, and z, we have $d(x) + d(y) + d(z) \geq m \geq n + 2$. Then G contains a cycle of length at least $\min\{n, 2m/3\}$.*

In [5], Bondy conjectured that the same result is true without the condition $m \geq n + 2$ and can be generalized to a k-connected graph for $k \geq 3$. In Theorem 7, below, we present a proof by Fournier and Fraisse [12] of this conjecture. Consequently Theorem 6 holds, and we do not need to present a proof of it.

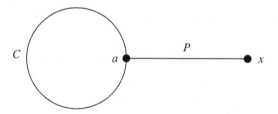

Fig. 2.4 *P* with only one point (an endpoint) on *C*

Suppose G is a k-connected graph on n vertices, $k \geq 2$, such that the degree-sum of any $k + 1$ independent vertices is at least m. We prove G contains a cycle of length at least $\min\{n, 2m/(k + 1)\}$. Several lemmas are involved.

We shall use the following notation:

1. Consider a cycle C in G for which an orientation has been chosen. If a and b are vertices of C, $[a, b]_C$ denotes the subgraph of C which is the path with endpoints a and b whose vertices are met on C when one goes on C from a to b.
2. For a vertex a of C, a^+ is the vertex on C immediately after a and (a^-) is the vertex just before a. Similarly, $a^{+\ell}$ $(a^{-\ell})$ is the ℓth vertex on C after (before) a. If S is a set of vertices of C, then $S^+ = \{a^+ \mid a \in S\}$.
3. If P is a path in G and if a and b are two vertices of P, $[a, b]_P$ denotes the subgraph of P which is the path with endpoints a and b.
4. If H is a subgraph of G and if a is a vertex of G, $\Gamma_H(a)$ is the set of neighbors of a in H, and $d_H(a) = |\Gamma_H(a)|$.
5. A C-path is a path of length at least two in G such that only its endpoints are on C.

Let C be a cycle of G of maximum length. We assume that C is not a Hamilton cycle. Let $R = G \setminus C$. Give an arbitrary orientation to C.

The proof of the following lemma can also be found in [9]:

Lemma 7.1. *Let P be a path of maximal length among all the paths with a given endpoint a on C and all the other vertices not on C (see Fig. 2.4). Let x be the other endpoint of P. Then there exists a C-path which contains x and all its neighbors in R such that one of its endpoints on C is a.*

Proof. If $d_R(x) = 0$ or 1, the result immediately follows from the fact that G is 2-connected. Assume $d_R(x) \geq 2$. Since P is of maximum length, all the neighbors of x in R are on P. Let x' be the furthest neighbor of x in R on P (see Fig. 2.5). Let C' be the cycle $[x, x']_P \cup \{xx'\}$. Let P_1 and P_2 be two disjoint paths between C and C' and let $Q = [x', a]_P$.

We now construct two disjoint paths P_3 and P_4 between C and C' such that x' is the endpoint of P_3 on C' and a is the endpoint of P_3 or P_4 on C.

Case 1. P_1 (respectively, P_2) does not intersect Q. Set $P_3 = Q$ and $P_4 = P_1$ (respectively, P_2).

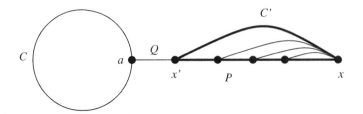

Fig. 2.5 x' is the furthest neighbor of x

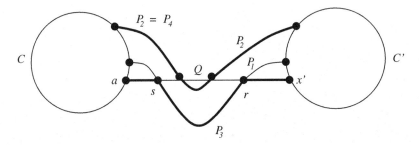

Fig. 2.6 r and s are on the same path

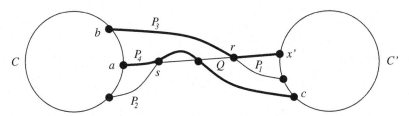

Fig. 2.7 r and s are on different paths

Case 2. P_1 and P_2 intersect Q. Let r be the vertex nearest to x' along Q and s be the vertex nearest to a along Q.

1. If r and s are on the same path P_1 (see, for example, Fig. 2.6), then

$$P_3 = [x', r]_Q \cup [r, s]_{P_1} \cup [s, a]_Q \text{ and } P_4 = P_2.$$

2. If $r \in P_1$ and $s \in P_2$, write b for the endpoint of P_1 on C and c for the endpoint of P_2 on C'. We set (see Fig. 2.7):

$$P_3 = [x', r]_Q \cup [r, b]_{P_1} \text{ and } P_4 = [c, s]_{P_2} \cup [s, a]_Q.$$

Let y be the endpoint of P_4 on C'. We orient C' so that $x \in [x', y]_{C'}$. We now have two possible cases:

1. If all the neighbors of x in P are on $[x', y]_{C'}$, we easily construct the C-path as required (see Fig. 2.8).

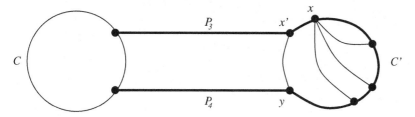

Fig. 2.8 The C-path in case 1

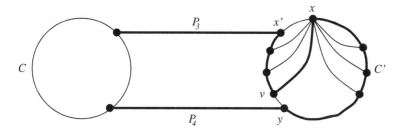

Fig. 2.9 The C-path in case 2

Fig. 2.10 The required new cycle

2. Suppose x has at least one neighbor on $[y^+, x'^-]_{C'}$. Let v be the neighbor of x on $[y^+, x'^-]_{C'}$ nearest y along C'. The C-path (see Fig. 2.9)

$$P_3 \cup [x', v]_{C'} \cup \{vx\} \cup [x, y]_{C'} \cup P_4$$

contains x and all its neighbors in R.   □

Lemma 7.2. *Suppose a, P, and x are defined as in Lemma 7.1. Then $d_R(x) \le \frac{|C|-4}{2}$.*

Proof. Use the C-path obtained in Lemma 7.1 to construct (see Fig. 2.10) a new cycle of length at least $\frac{|C|+2}{2} + d_R(x) + 1$. Hence, $(|C| + 2)/2 + d_R(x) + 1 \le |C|$ and the result follows.   □

Lemma 7.3. *Let a, P, and x be as in Lemma 7.1. Then $d(x) \le |C|/2$.*

Proof. Since the cycle C is of maximum length, it follows that x is not adjacent to the vertex $a^{+\ell}$ if $1 \leq \ell \leq \min\{|P| - 1, |C| - 1\}$. We now have

1. $|P| - 1 \geq d_R(x) + 1$ by the choice of P;
2. $|C| - 1 \geq 2d_R(x) + 3$ by Lemma 7.2.

Thus x is not adjacent to $a^{+\ell}$ for $1 \leq \ell \leq d_R(x) + 1$. Similarly, x is not adjacent to $a^{-\ell}$ for $1 \leq \ell \leq d_R(x) + 1$.

Now by Lemma 7.2

$$2(d_R(x) + 1) \leq 2\left(\frac{|C| - 4}{2} + 1\right) \leq |C| - 2.$$

Therefore the set of vertices $\{a^{+\ell} \mid 1 \leq \ell \leq d_R(x) + 1\}$ cannot intersect the set of vertices $\{a^{-\ell} \mid 1 \leq \ell \leq d_R(x) + 1\}$. In addition, since C is of maximum length, it follows that no two consecutive vertices of C are simultaneously adjacent to x. Hence,

$$d_C(x) \leq 1 + \frac{|C| - 1 - 2(d_R(x) + 1) + 1}{2} \leq \frac{|C|}{2} - d_R(x).$$

Therefore $d(x) \leq |C|/2$, as required.   □

Lemma 7.4. *For any C-path with endpoints a_1 and a_2 on C, which is not reduced to an edge,*

$$d_C(a_1^+) + d_C(a_2^+) \leq |C|.$$

Proof. Since C is of maximal length,

$$\Gamma_C^+(a_1^+) \cap \Gamma_C(a_2^+) \cap [a_2^{++}, a_1] = \emptyset$$

(see Fig. 2.11). Thus

(1) $d_{[a_2^+, a_1^-]}(a_1^+) + d_{[a_2^{++}, a_1]}(a_2^+) \leq |[a_2^{++}, a_1]|.$

Similarly,

(2) $d_{[a_1^{++}, a_2]}(a_1^+) + d_{[a_1^+, a_2^-]}(a_2^+) \leq |[a_1^{++}, a_2]|.$

Now add the inequalities (1) and (2) to obtain $d_C(a_1^+) + d_C(a_2^+) \leq |C|$.   □

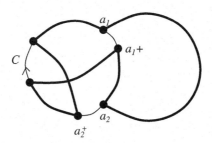

Fig. 2.11 A C-path with a_1 and a_2 on C

Lemma 7.5. *Define a, P, and x as in Lemma 7.1. Then $d(x) + d_C(a^+) \leq |C|$.*

Proof. We consider two cases.

Case 1. $d_{C \setminus \{a\}}(x) = 0$. By Lemma 7.1, there exists a C-path which contains x and all its neighbors in R such that one of its endpoints on C is a. Let b be the other endpoint of this C-path. The vertex a^+ is not adjacent to $b^{+\ell}$ if $1 \leq \ell \leq d_R(x) + 1$ because C is of maximum length (see Fig. 2.12). Therefore $d_C(a^+) \leq |C| - (d_R(x) + 1)$, that is, $d_C(a^+) + d_R(x) \leq |C| - 1$. Now since $d_C(x) \leq 1$, it follows that $d_C(a^+) + d(x) \leq |C|$.

Case 2. $d_{C \setminus \{a\}}(x) \geq 1$. Let b be the neighbor of x on $C \setminus \{a\}$ such that $d_{[b^+, a^-]_C}(x) = 0$. Then $|[b^+, a^-]_C| \geq d_R(x) + 1$. In addition, a^+ is not adjacent to any of the vertices $b^{+\ell}$ for $1 \leq \ell \leq d_R(x) + 1$ (see Fig. 2.13).

On the other hand, let v be any neighbor of x on C, different from a and b. Then $v \notin [b, a]_C$. So $v^+ \notin \{b^{+\ell} \mid 1 \leq \ell \leq d_R(x) + 1\}$ and a^+ is not adjacent to v^+ (see Fig. 2.14).

Finally, since a^+ is not adjacent to itself,

$$d_C(a^+) \leq |C| - 1 - (d_R(x) + 1) - (d_C(x) - 2)$$
$$\leq |C| - d_R(x) - d_C(x).$$

Therefore $d(x) + d_C(a^+) \leq |C|$, as required.   □

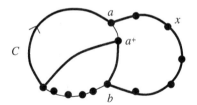

Fig. 2.12 Case 1: $d_{C \setminus \{a\}}(x) = 0$

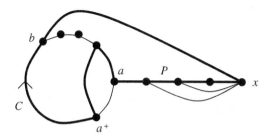

Fig. 2.13 Case 2: $d_{C \setminus \{a\}}(x) \geq 1$ and a^+ is not adjacent to $b^{+\ell}$

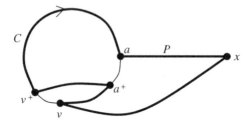

Fig. 2.14 Case 2: $d_{C\setminus\{a\}}(x) \geq 1$ and a^+ is not adjacent to v^+

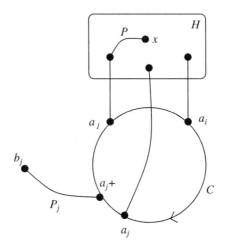

Fig. 2.15 k parallel edges between H and C

We are now ready to prove the following theorem of Fournier and Fraisse [12]:

Theorem 7. *Let G be a k-connected graph where $k \geq 2$, such that the degree-sum of any $k + 1$ independent vertices is at least m. Then G contains a cycle of length at least* $\min\{n, 2m/(k + 1)\}$.

Proof. Let a_1 be a vertex of C such that $d_R(a_1^+)$ is minimum. Suppose that P is a path with an endpoint a_1 such that $V(P) \setminus \{a_1\} \subset R$ and of maximum length. Let x be the other endpoint of P in R and let H be the component of x in R.

Since G is k-connected, there exist k edges between H and C with distinct endpoints on C and we can suppose that one of these edges is the edge of P containing a_1 as an endpoint. Note that these edges do not necessarily have distinct endpoints in H. We denote the endpoints of these edges on C by $a_1, a_2, \ldots, a_k$ (see Fig. 2.15).

Define

$$I = \{i \in \{1, 2, \ldots, k\} \mid d_R(a_i^+) = 0\}$$

and

$$J = \{j \in \{1, 2, \ldots, k\} \mid d_R(a_j^+) \neq 0\}.$$

Let $P_j, j \in J$, be a path starting from a_j^+ such that $v(P_j) \setminus \{a_j^*\} \subset R$ and of maximum length. Assume b_j is the endpoint of P_j in R. Then

1. P_j does not intersect H (and P) because there is no C-path between a_j and a_j^+.
2. If $j_1 \neq j_2$, then P_{j_1} does not intersect P_{j_2} because it is impossible to have a C-path between a_{j_1} and a_{j_2} through H and also a distinct C-path between $a_{j_1}^+$ and $a_{j_2}^+$.
3. The set $\{x\} \cup \{a_i^+ \mid i \in I\} \cup \{b_j \mid j \in J\}$ is an independent set.
4. $d(x) \leq |C|/2$ by Lemma 7.3.
5. $d(b_j) \leq |C|/2$ for $j \in J$ by Lemma 7.3.

We now consider three cases.

Case 1. $|I| \geq 2$. Let $p = |I|$ and $I = \{i_1, i_2, \ldots, i_p\}$. Then $d_R(a_{i_1}^+) = d_R(a_{i_2}^+) = \ldots = d_R(a_{i_p}^+) = 0$. By Lemma 7.4,

$$\begin{aligned} d_C(a_{i_1}^+) + d_C(a_{i_2}^+) &\leq |C| \\ d_C(a_{i_2}^+) + d_C(a_{i_3}^+) &\leq |C| \end{aligned}$$

$$\cdot$$
$$\cdot$$
$$\cdot$$

$$\begin{aligned} d_C(a_{i_{p-1}}^+) + d_C(a_{i_p}^+) &\leq |C| \\ d_C(a_{i_p}^+) + d_C(a_{i_1}^+) &\leq |C|, \end{aligned}$$

so $\sum_{i \in I} d(a_i^+) \leq p(|C|/2)$. Therefore

$$m \leq d(x) + \sum_{i \in I} d(a_i^+) + \sum_{j \in J} d(b_j) \leq (k+1)\frac{|C|}{2}.$$

Hence, $|C| \geq \frac{2m}{k+1}$, as required.

Case 2. $|I| = 1$. From the choice of a_1 we must have $I = \{1\}$. Now by Lemma 7.5, $d_C(a_i^+) + d(x) \leq |C|$. In addition, since $d_R(a_i^+) = 0$, it follows that

$$m \leq d(x) + d(a_1^+) + \sum_{j=2}^{k} d(b_j) \leq (k+1)\frac{|C|}{2}.$$

Hence, $|C| \geq \frac{2m}{k+1}$.

Case 3. $|I| = 0$. In this case we have

$$m \leq d(x) + \sum_{j=1}^{k} d(b_j) \leq (k+1)\frac{|C|}{2}.$$

Hence, $|C| \geq \frac{2m}{k+1}$. This completes the proof.   □

The following theorem is an immediate result of Theorem 7 with $k = 2$:

Theorem 8 ([5]). *Let G be a 2-connected graph on n vertices. Suppose that for every set of three independent vertices x, y, and z, we have $d(x) + d(y) + d(z) \geq 3n/2 - 1$. Then G is Hamiltonian.*

Chapter 3
Pancyclicity

3.1 Introduction

Recall the definition given in Sect. 1.4: we define a *pancyclic* graph to be a graph on $n \geq 3$ vertices containing a cycle of every length from 3 to n inclusive. The existence of such graphs is not in question—the complete graph is obviously pancyclic—and other examples are easy to construct. For example, if one takes the n-cycle $(v_1, v_2, \ldots v_n, v_1)$ and adds an edge from v_1 to v_i, the resulting graph contains cycles of lengths i and $n - i + 2$, namely $(v_1, v_2, \ldots v_i, v_1)$ and $(v_1, v_i, v_{i+1}, \ldots v_n, v_1)$, so the addition of edges (v_1, v_i) for $i = 3, 4, \ldots, \lfloor \frac{n}{2} \rfloor$ constructs a pancyclic graph.

Two kinds of questions have historically been asked concerning pancyclic graphs. First, what is the minimum number of edges, or what are the degree properties, required in order to guarantee that all graphs satisfying those conditions are pancyclic; in particular, how much stronger must a sufficient condition for the Hamiltonian property be to guarantee the pancyclic property? Second, what are the smallest pancyclic graphs for a given value of n?

Questions of this second kind ask for precise bounds on some function $m(n)$ so that no graph on n vertices with fewer than $n + m(n)$ edges will be pancyclic. Such a function clearly exists; for instance, no graph on n vertices with $n - 1$ or fewer edges can be Hamilton, and therefore cannot be pancyclic. Chapter 4 will deal with this question. In this chapter, we look at questions of the first kind.

© The Author(s) 2016
J.C. George et al., *Pancyclic and Bipancyclic Graphs*, SpringerBriefs
in Mathematics, DOI 10.1007/978-3-319-31951-3_3

3.2 Bounds

We state the most commonly cited sufficient conditions for the Hamiltonian property. Historically the first such condition was Dirac's theorem:

Theorem 9 ([9]). *If each vertex of the graph G has degree at least $n/2$ and $n \geq 3$, then G has a Hamilton cycle.*

Rather than proving Dirac's theorem itself, we usually prove it as an easy corollary to the theorem of Ore:

Theorem 10 ([25]). *If any pair x and y of non-adjacent vertices of G satisfy $d(x) + d(y) \geq n$, then G has a Hamilton cycle.*

This theorem may be proven on its own, but it also follows from Theorem 4 of Chvátal stated in the previous chapter:

G is Hamiltonian if its degree sequence $d_1 \leq d_2 \leq \cdots \leq d_v$ satisfies the condition that for $i < v/2$ we have either $d_i > i$ or $d_{v-i} > v - i$.

For the sake of brevity, we will call this the *Chvátal condition*. Notice also that, in general, $d_i \neq d(i)$ because we will order the vertices not by degree but in order along a Hamilton cycle.

The following fundamental result is the basis of many results that tie sufficient conditions for the Hamilton property to the pancyclic property. It is found in [3].

Theorem 11. *If G is Hamiltonian and $e \geq n^2/4$, then either G is the complete bipartite graph $K_{n/2,n/2}$ or G is pancyclic.*

Corollary 11.1. *If G satisfies the conditions of Ore's Theorem, then either G is pancyclic or G is $K_{n/2,n/2}$.*

In [4], Bondy stated the "metaconjecture," which says that if G satisfies the conditions of almost any nontrivial theorem that implies that G is Hamiltonian, then G is pancyclic, with possibly a small family of exceptions (such as the complete bipartite graphs). This was inspired by the corollary above, which confirms the metaconjecture for the theorems of Ore and of Dirac (whose theorem is implied by that of Ore). The preceding theorem and corollary will follow from the proof of a more general theorem below.

Our goal is to show that the three theorems of the previous chapter that imply the Hamilton property also imply the pancyclic property, with the exception of at most a small family of counterexamples. These results come from [2], and we follow their proofs. We assume that the graph G has vertices $\{v_i : 1 \leq i \leq n\}$, where $(v_1, v_2, \ldots, v_n, v_1)$ is a Hamilton cycle whose existence follows from one of the Theorems 4, 5, or 8.

Theorem 12 (Schmeichel and Hakimi [28]). *A graph that satisfies the Chvátal condition is either pancyclic or bipartite.*

Proof. The proof of the theorem employs a sequence of lemmas.

Lemma 12.1. *Let G be a graph on n vertices, and suppose G contains an* $(n-1)$-*cycle that omits the vertex x. If* $d(x) \geq n/2$, *then G is pancyclic.*

Proof. Let the cycle be $(v_1, v_2, \ldots, v_{n-1}, v_1)$. Let $e(i, j)$ be 1 if v_i and v_j are adjacent, and 0 otherwise. If G contains no cycle of length ℓ, then $e(n, i) + e(n, i + \ell - 2) \leq 1$ for each $1 \leq i \leq n - 1$ (where we reduce i modulo $n - 1$ as needed). Now we see

$$d(v_n) = \frac{1}{2} \sum_{i=1}^{n-1} [e(n, i) + e(n, i + \ell - 2)] \leq (n-1)/2 < n/2.$$

Thus if $d(v_n) \geq n/2$ there must be a cycle of each length ℓ. □

Lemma 12.2 (Bondy [3]). *Let G have a Hamilton cycle* $(v_1, v_2, \ldots, v_n, v_1)$. *If for some i with* $1 \leq i \leq n$ *we have* $d(v_i) + d(v_{i+1}) > n$, *then G is pancyclic (where we identify* v_{n+1} *with* v_1).

Proof. Suppose that G has no cycle of length ℓ for some $3 \leq \ell < n$. For any vertices i and $i + 1$ define for $k \notin \{i, i+1\}$

$$f_\ell(k) = \begin{cases} k - \ell + 1, j + 2 \leq k \leq i + \ell - 2 \\ k - \ell + 3, i + \ell - 1 \leq k \leq i - 1 \end{cases}$$

We prove that if G contains no cycle of length ℓ, then for any i with $1 \leq i \leq n$, $d(v_i) + d(v_{i+1}) \leq n$. Equality holds if and only if for each value of $k \notin \{i, i+1\}$ exactly one of the edges (i, k) or $(i + 1, f_\ell(k))$ is in G.

If G includes both these edges then they, the edge $(i, i+1)$, and the path between k and $f_\ell(k)$ along the cycle that does *not* pass through i, together constitute a cycle of length ℓ. If there is no such cycle, then for each i at least one of those two edges is missing. Now $f_\ell(k)$ is one-to-one. Thus, of the $2n - 4$ possible edges that connect the set $\{v_i, v_{i+1}\}$ with the rest of the graph, at most half may be in the graph, as each edge from v_i to v_k eliminates some edge incident to v_{i+1}. □

This ability to exclude one of two edges in a Hamiltonian graph with no cycle of length ℓ is a useful technique. Following Schmeichel and Hakimi [28], we refer to this as the ℓ-*correspondence principle*.

Lemma 12.3. *Suppose G contains a Hamilton cycle* $(v_1, v_2, \ldots, v_n, v_1)$ *and for some i we have* $d(v_i) + d(v_{i+1}) \geq n$. *We may assume that* $d(v_{i+1}) \leq d(v_i)$ *(if not, reverse the ordering of the subscripts). If G is neither pancyclic nor bipartite, then the following are true:*

1. *G has cycles of each length other than* $n - 1$;
2. *There are no edges among* $\{v_{i-2}, v_{i-1}, v_i, v_{i+1}, v_{i+2}, v_{i+3}\}$ *except the edges of the cycle;*
3. *Each of the vertices* $\{v_{i-2}, v_{i-1}, v_i, v_{i+1}, v_{i+2}, v_{i+3}\}$ *has degree less than* $n/2$, *so that at most half the vertices of G have degree* $n/2$ *or higher; and*
4. *If* $d(v_i) = d(v_{i+1}) = n/2$, *then G contains the edges* $v_i v_{i-4}$, $v_i v_{i-3}$, $v_{i+1} v_{i+4}$, *and* $v_{i+1} v_{i+5}$.

Proof. We begin by renumbering the vertices of G so that the two vertices v_i, v_{i+1} are v_n and v_1 respectively, and $d(v_1) \leq d(v_n)$. We separate the proof into several lemmas.

Lemma 12.4. *Let G have a Hamilton cycle $(v_1, v_2, \ldots, v_n, v_1)$. If G is not bipartite and not pancyclic, then for some j G contains a 3-cycle (v_i, v_j, v_{j+1}, v_i).*

Proof. We seek to find a 3-cycle (v_1, v_j, v_{j+1}, v_1).

If $d(v_1) > n/2$, then clearly v_1 has two neighbors that are adjacent on the cycle and we are done. Since $d(v_1) + d(v_n) = n$ and $d(v_1) \leq d(v_n)$, we must have $d(v_n) = d(v_1) = n/2$. It follows that n is even. Then if v_n is not to be adjacent to two consecutive vertices in the cycle, it must be adjacent to all vertices of odd subscript (and only to those vertices). It follows that G contains a cycle of every even length containing $v_n v_1$. There are two cases to be considered.

Case 1. G contains a 3-cycle (v_1, v_j, v_{j+1}, v_1) for some j. If $j = 2$ or $j = n - 1$, G would be pancyclic; cycles of every odd length would be formed by taking the cycles of every even length and replacing the edge $v_n v_1$ with the other two edges of the 3-cycle. Then $3 \leq j \leq n - 2$. Now to construct odd cycles of length $\ell \geq 5$, we choose a vertex v_s where $s \leq j < j + 1 \leq s + \ell - 3 \leq n - 1$. The even cycle $(v_n, v_s, v_{s+1}, \ldots, v_{s+\ell-3}, v_n)$ contains the edge $v_j v_{j+1}$; replacing this edge by the edges $v_j v_1$ and $v_1 v_{j+1}$ gives us the required odd cycle, contradicting the hypothesis.

Case 2. G has no such 3-cycle. Since $d(v_1) = n/2$, v_1 must be adjacent to all vertices of even subscript (and only to those vertices). As G is not bipartite by assumption, there must be an edge connecting two vertices with subscripts of the same parity, or else the vertices of even subscript and those of odd subscript would form a bipartition. This edge, together with the edges of the cycle and the edges incident with v_1 and v_n, gives us cycles of every odd length, contradicting the hypothesis.

□

Corollary 12.1. *If G has a Hamilton cycle but is neither pancyclic nor bipartite, and two consecutive vertices v_n, v_1 on the cycle each have degree $n/2$, then G contains two 3-cycles (v_1, v_i, v_{i+1}, v_1) and (v_n, v_j, v_{j+1}, v_n).*

Lemma 12.5. *If G has a Hamilton cycle, is neither pancyclic nor bipartite, and $d(v_1) = d(v_n) = n/2$, then G contains a 3-cycle (v_1, v_k, v_n, v_1) for some k.*

Proof. By Lemma 12.4, we have the 3-cycle (v_n, v_i, v_{i+1}, v_n) for some i. Assume there is no 3-cycle (v_1, v_k, v_n, v_1); then v_n and v_1 have no common neighbor v_k. It follows, since $d(v_n) + d(v_1) = n$, that each vertex of G is adjacent to exactly one of v_1 and v_n. We use this to construct an ℓ-cycle for each ℓ, $3 \leq \ell \leq n$, which is a contradiction.

Case 1. If $\ell \leq i + 1$, then $2 \leq i - \ell + 3 \leq i - 1$; consider the vertex $v_{i-\ell+3}$. It must be adjacent to one of v_1 or v_n. If it is adjacent to v_1, then $(v_n, v_i, v_{i-1}, \ldots, v_{i-\ell+3}, v_1, v_n)$ is the ℓ-cycle. If it is adjacent to v_n, then $(v_n, v_{i+1}, v_i, \ldots, v_{i-\ell+3}, v_n)$ is the ℓ-cycle.

Case 2. If $\ell \geq i + 2$, then consider $v_{\ell-1}$. If this vertex is adjacent to v_1, then $(v_1, v_2, \ldots, v_i, v_n, v_{i+1}, v_{i+2}, \ldots, v_{\ell-1}, v_1)$ is an ℓ-cycle; if instead it is adjacent to v_n, then $(v_1, v_2, \ldots, v_{\ell-1}, v_n, v_1)$ is the ℓ-cycle. □

Lemma 12.6. *If G satisfies the hypotheses of Lemma 12.5, then G cannot contain either of the edges $v_1 v_{n-1}$ or $v_n v_2$.*

Proof. We suppose G contains either of these edges and demonstrate that G must be pancyclic, establishing a contradiction.

Case 1. Suppose first G contains the edge $v_1 v_{n-1}$. As G clearly contains cycles of length 3, n, and $n - 1$, we need merely construct cycles of length ℓ where $4 \leq \ell \leq n - 2$.

If the edge $v_1 v_{n-\ell+1} \in G$, we have our ℓ-cycle $(v_1, v_{n-\ell+1}, v_{n-\ell+2}, \ldots, v_{n-1}, v_1)$; so by the ℓ-correspondence principle enunciated after Lemma 12.2, G must contain the edge $v_n v_{n-2}$. Also, if $v_n v_{\ell-2}$ is an edge of G, we have the ℓ-cycle $(v_1, v_2, \ldots v_{\ell-2}, v_n, v_{n-1}, v_1)$. Similarly, $v_n v_{\ell-1}$ gives us $(v_1, v_2, \ldots v_{\ell-1}, v_n, v_1)$. Accordingly, we look at the case in which these edges are not present.

Suppose that for some i we have $v_n v_i$ in G but $v_n v_{i+1}$ not in G, where $\ell \leq i \leq n - 2$. The ℓ-correspondence principle would require the edge $v_1 v_{i-\ell+4}$ to be in G; then there would be an ℓ-cycle $(v_1, v_{i-\ell+4}, v_{i-\ell+5}, \ldots v_i, v_n, v_{n-1}, v_1)$ in G. Thus G must contain all edges $v_n v_{n-1}, v_n v_{n-2}, \ldots v_n v_s$ where $\ell \leq s \leq n - 2$, and contain none of the edges $v_n v_{s-1}, v_n v_{s-2}, \ldots v_n v_{\ell-2}$.

Similarly, suppose $v_n v_i$ is in G but $v_n v_{i-1}$ is not in G, for some $i \leq \ell - 3$. Then the ℓ-correspondence principle guarantees that the edge $v_1 v_{n-\ell+i}$ is in G, and $n - \ell + i \leq n - 3$. We would then have an ℓ-cycle

$$(v_1, v_2, \ldots, v_i, v_n, v_{n-2}, v_{n-3}, \ldots, v_{n-\ell+i}, 1).$$

From this we can see that the neighbors of v_n are the vertices $v_1, v_2, \ldots v_r$ and $v_s, v_{s+1}, \ldots, v_{n-1}$ where r and s satisfy $s - r = n - d(v_n)$. But we will see that these edges can be used to construct a cycle of any desired length, which contradicts the assumption that G is not pancyclic. So cycles of length at least $n - d(n) + 2$ may be formed by starting with the cycle $(v_n, v_r, v_{r+1}, \ldots, v_s, v_n)$ and lengthening it as required by inserting the vertices v_{s+1}, v_{s+2}, and so on, between v_s and v_n. Cycles of length no greater than $n - d(n) + 1$ may be formed by starting with the cycle $(v_s, v_s, v_{s+1}, \ldots, v_{n-1}, v_1, v_2, \ldots v_r, v_n)$ and removing vertices v_r, v_{r-1}, and so on.

Case 2. Suppose G contains $v_n v_2$. Following the argument for **Case 1**, we show that the neighbors of v_1 are v_2, v_3, up to v_r, and v_s, v_{s+1}, up through v_n, where $s - r = n - d(v_1)$. We find a cycle $(v_2, v_3, \ldots, v_r, v_1, v_s, v_{s+1}, \ldots, v_n, v_2)$. This cycle has length $d(v_1) + 1$, and we can shorten it, so G contains cycles of every length up to $d(v_1) + 1$. We use the ℓ-correspondence principle to show that the neighbors of v_n must be v_1 through v_p and v_q through v_{n-1}, where $q - p = n - d(v_n)$. Now G contains the cycle $(v_n, v_p, v_{p+1}, \ldots, v_q, v_n)$. This cycle has length $d(v_1) + 2$, and we can get a cycle of any longer length by inserting vertices v_{p-1}, v_{p-2}, and so forth between v_n and v_p. It follows that G is pancyclic. □

To finish the proof of Lemma 12.3, suppose G is not bipartite, and G has no cycle of length ℓ, where $4 \leq \ell \leq n - 1$. By Lemma 12.6, G has neither the edge $v_1 v_{n-1}$ nor $v_n v_2$. It follows that G contains the edges $v_1 v_{n-\ell+3}$ and $v_n v_{\ell-2}$.

Lemma 12.7. *Under the hypotheses of Lemma 12.3, $\ell > n/2 + 2$.*

Proof. Consider the 3-cycle (v_1, v_k, v_n, v_1) guaranteed by Lemma 12.5. If $k > \ell - 2$, then we have an ℓ-cycle $(v_1, v_2, \ldots, v_{\ell-2}, v_n, v_k, v_1)$ using two of the edges of that 3-cycle; so $k \leq \ell - 2$. Similarly, if $k < n - \ell + 3$, we get an ℓ-cycle $(v_1, v_{n-\ell+3}, v_{n-\ell+4}, \ldots, v_n, v_k, v_1)$. Thus $n - \ell + 3 \leq k \leq \ell - 2$. This quickly gives us $2\ell > n + 5$, so $\ell > n/2 + 2$. $\qquad\square$

Lemma 12.8. *G cannot contain both the edges $v_1 v_{\ell-1}$ and $v_n v_{n-\ell+2}$.*

Proof. Suppose both these edges were in G. Then if for some i between 1 and $\ell - 2$ we have $v_n v_i$ and $v_n v_{i+1}$ are both edges of G, we would have the ℓ-cycle $(v_1, v_2, \ldots, v_i, v_n, v_{i+1}, v_{i+2}, \ldots, v_{\ell-1}, v_1)$. Similarly, if neither of these edges $v_1 v_{\ell-1}$ and $v_n v_{n-\ell+2}$ are in G, then by the ℓ-correspondence principle we would have both the edges $v_1 v_{n-\ell+i+1}$ and $v_1 v_{n-\ell+i+2}$ in G. These edges would give us the ℓ-cycle $(v_n, v_{n-\ell+2}, \ldots, v_{n-\ell+i+1}, v_1, v_{n-\ell+i+2}, \ldots, v_n)$.

It follows that the presence of the two edges $v_1 v_{\ell-1}$ and $v_n v_{n-\ell+2}$ in G would require that the neighbors v_i for $i \leq \ell-2$ of v_n are precisely the vertices $\{v_i : i \text{ odd}\}$. In much the same way, we can show that the neighbors v_j for $j \geq n - \ell + 3$ of v_1 in G would be precisely $\{v_j : j \text{ even}\}$. Now, Lemma 12.5 guarantees the existence of a 3-cycle (v_1, v_k, v_n, v_1); but this common neighbor v_k cannot by the foregoing argument satisfy $k \leq \ell - 2$ or $k \geq n - \ell + 3$. We can then construct an ℓ-cycle as we did in the proof that $\ell > n/2 + 2$. We now conclude that at most one of the two edges $v_1 v_{\ell-1}$ and $v_n v_{n-\ell+2}$ is in G. $\qquad\square$

From here, we assume that G does not contain $v_1 v_{\ell-1}$. The proof in the other case is similar.

Lemma 12.9. *G contains none of the edges $\{v_1 v_i : \ell - 1 \leq i \leq n - 1\}$.*

Proof. If $\ell = n - 1$, this is immediate, so we assume $\ell \leq n - 2$. The absence of $v_1 v_{\ell-1}$ from G implies by the ℓ-correspondence principle that G contains the edge $v_n v_{2\ell-n-2}$. Now, G does not contain the edge $v_1 v_{n-1}$ by Lemma 12.6; suppose v_1 has a neighbor v_j where $\ell \leq j \leq n-2$. Now we look for the least i with $j \leq i \leq n-2$ for which v_i is adjacent to v_1 but v_{i+1} is not adjacent to v_1. Such an i must exist since v_j is adjacent to v_1 and v_{n-1} is not.

Because the edge $v_1 v_{i+1}$ is not in G, the ℓ-correspondence principle requires that $v_n v_{i+\ell-n}$ is in G. This produces an ℓ-cycle

$$(v_1, v_2, \ldots, v_{2\ell-n-2}, v_n, v_{i+\ell-n}, v_{i+\ell-n+1}, \ldots, v_i, v_1).$$

Hence G contains none of the edges $\{v_1 v_i : \ell - 1 \leq i \leq n - 1\}$. $\qquad\square$

Lemma 12.10. *If v_n is adjacent to any vertex v_i where $i \geq \ell$, then v_n must be adjacent to each vertex $v_{i+1}, v_{i+2}, \ldots v_{n-1}$.*

Proof. Suppose then $v_n v_i$ is in G but $v_n v_{i+m}$ is not, where $i + m \leq n - 1$. We consider the graph H that results when vertices $v_{i+1}, v_{i+2}, \ldots v_{n-1}$ are deleted from G. Notice that H has $i + 1$ vertices and a Hamilton cycle made from the edges of the Hamilton cycle of G together with the edge $v_i v_n$.

We know that G does not contain edges $v_n v_{i+m}$ and $v_1 v_j$ for $i + 1 \leq j \leq n - 1$; thus H must contain all the neighbors of v_1. It follows that the degrees in H of these two vertices must add to at least $|H| > \ell$. Now, by Lemma 12.2, H is pancyclic, and therefore has a cycle of length ℓ; thus G also has a cycle of length ℓ. It follows that G is pancyclic; from this we conclude that if v_n is adjacent to any vertex v_i where $i \geq \ell$, then v_n must be adjacent to each vertex $v_{i+1}, v_{i+2}, \ldots v_{n-1}$. □

Lemma 12.11. *G contains cycles of all lengths except $\ell = n - 1$.*

Proof. We know that $v_1 v_{n-\ell+3}$ is in G, and $v_1 v_{\ell-1}$ is not in G; thus there is a first subscript i between $n - \ell + 3$ and $\ell - 1$ for which v_i is adjacent to v_1 but v_{i+1} is not. The ℓ-correspondence principle thus guarantees that the edge $v_n v_{i+\ell-n}$ is in G, and $i + \ell - n > 2$. This gives us a cycle of length $\ell + 1$, $(v_1, v_2, \ldots, v_{i+\ell-n}, v_n, v_{n-1}, \ldots, v_i, v_1)$.

Suppose $v_n v_{n-2}$ is in G; then we can construct an ℓ-cycle by replacing the edges $v_n v_{n-1}$ and $v_{n-1} v_{n-2}$ by this edge. But if $v_n v_{n-2}$ is not an edge in G and $\ell < n - 1$ then G does not contain the edge $v_n v_\ell$ by Lemma 12.10. Now the ℓ-correspondence principle says that the edge $v_1 v_3$ is in G; we find an ℓ-cycle by replacing the edges $v_1 v_2$ and $v_2 v_3$ with $v_1 v_3$ to get our ℓ-cycle. This means $\ell = n - 1$ is the only remaining possibility. □

This completes the proof of Lemma 12.3, item 1. If G has no cycle of length $n-1$, then G cannot contain the edges $v_1 v_{n-1}$ or $v_n v_2$; the ℓ-correspondence principle, with $\ell = n - 1$, guarantees that G contains the edges $v_1 v_4$ and $v_n v_{n-3}$.

We claim that G does not contain the edge $v_1 v_{n-2}$; if it did, we employ the 3-cycle (v_n, v_i, v_{i+1}, v_n) whose existence follows from Lemma 12.4 to create an $(n - 1)$-cycle $(v_1, v_2, \ldots, v_i, v_n, v_{i+1}, v_{i+2}, \ldots, v_{n-2}, v_1)$. Since G does not contain $v_1 v_{n-2}$, by the ℓ-correspondence principle it must contain $v_n v_{n-4}$.

Now G cannot contain the edges $v_2 v_{n-1}$, $v_2 v_{n-2}$, $v_3 v_{n-1}$, or $v_3 v_{n-2}$. We give the $(n - 1)$-cycle $(v_2, \ldots, v_{n-4}, v_n, v_{n-3}, v_{n-2}, v_{n-1}, v_2)$ produced by the edge $v_2 v_{n-1}$; the other edges produce $(n - 1)$-cycles similarly.

This completes the proof of Lemma 12.3, item 2.

Now we show that the vertices v_{n-2}, v_{n-1}, v_2, and v_3 each have degree less than $n/2$. Note that, by the $(n - 1)$-correspondence principle exactly one of the two edges $v_n v_3$ and $v_1 v_5$ must be in G. Assume that $v_1 v_5$ is an edge of G; the proof for the other case is similar. Let $S = \{v_1, v_2, v_3, v_n, v_{n-1}, v_{n-2}, v_{n-3}, v_{n-4}\}$. The only neighbors in S of v_{n-2} are v_{n-1} and v_{n-3}. Thus if v_{n-2} were to have degree $n/2 - 1$ or greater, it would have two consecutive neighbors v_i and v_{i+1} among the vertices not in S. This would give us the $(n - 1)$-cycle

$$(v_1, v_2, \ldots, v_i, v_{n-2}, v_{i+1}, v_{i+2}, \ldots, v_5, v_1).$$

Essentially the same line of reasoning shows that the degree of v_3 is less than $n/2-1$; for the other vertices, let $T = \{v_1, v_2, v_3, v_4, v_n, v_{n-1}, v_{n-2}, v_{n-3}\}$. If $v_4 v_{n-1}$ were an edge of G, we would have an $(n-1)$-cycle

$$(v_1, v_2, v_3, v_4, v_{n-1}, v_{n-2}, \ldots, v_5, v_1).$$

Thus v_{n-1} has only two neighbors v_n and v_{n-2} in T, and as before if $d(v_{n-1}) \geq n/2$ there would be consecutive vertices v_j, v_{j+1} adjacent to v_{n-1} that would provide an $(n-1)$-cycle

$$(v_1, v_2, \ldots, v_i, v_{n-1}, v_{i+1}, v_{i+2}, \ldots, v_{n+3}, v_n, v_1).$$

If instead the edge $v_n v_3$ is in G, the degrees of those two vertices would be less than $n/2$ rather than $n/2 - 1$. Also, it may be that $v_{n-1} v_4$ is in G, so we could only conclude that the degree of $v_{n-1} < n/2$. This completes the proof of Lemma 12.3, items 3 and 4.   □

We are now in a position to prove Theorem 12. First, we show that it is true in nearly all cases with yet another lemma.

Lemma 12.12. *Let G satisfy the Chvátal condition. If n is odd, or if we have both n even and $d_{n/2} \neq n/2$, then G is pancyclic.*

Proof. Suppose first that n is odd and consider $d_{(n-1)/2}$. If this quantity is greater than $n/2$, then $d_{(n+1)/2} > n/2$ as well. Contrarily, if $d_{(n-1)/2} < n/2$, then by the Chvátal condition, $d_{(n+1)/2} > n/2$. Hence, more than half the vertices have degree $(n+1)/2$, so two such vertices must be consecutive along the Hamilton cycle. Then by Lemma 12.2 the graph is pancyclic.

In the case that n is even, suppose that $d_{n/2} > n/2$. Then as above more than half the vertices have degree greater than $n/2$, so G is pancyclic. On the other hand, if $d_{n/2} < n/2$, then $d_{n/2-1} < n/2$ and the Chvátal condition guarantees that $d_{n/2+1} \geq n/2+1 > n/2$. Thus exactly half the vertices have degree greater than $n/2$. If two of these are adjacent along the cycle, we may conclude that G is pancyclic; otherwise, without loss of generality, we may assume that the odd-numbered vertices are those with degree greater than $n/2$.

If $v_1 \sim v_{n-1}$ in G, consider the graph G' formed by deleting the vertex v_n; the edge $v_1 v_{n-1}$ together with the remaining edges of the Hamilton cycle of G forms a Hamilton cycle of G', and more than half its vertices have degree greater than $n/2$. It follows that G', and therefore G, is pancyclic. We are left with the case in which n is even and v_1 is not adjacent to v_{n-1}. Now, G contains a 3-cycle as any vertex with degree greater than $n/2$ must be a common neighbor of two vertices that are adjacent in the Hamilton cycle. So assume that G contains no cycle of length $\ell \geq 4$ and consider the neighbors of vertex v_{n-1}. Since $n-1$ is odd, there are more than $n/2$ such neighbors, and v_1 is not among them. Each edge $v_{n-1} v_k$ obliges us to exclude from G an edge $v_1 v_{g_\ell(k)}$ with which it would form a cycle of length ℓ; in this case

$$g_\ell(k) = \begin{cases} k - \ell, & 1 \leq k \leq \ell - 4 \\ k - \ell + 4, & \ell - 3 \leq k \leq v - 4 \end{cases}$$

The vertex v_{n-1} is not adjacent to v_1, and has degree greater than $n/2$ since $n-1$ is odd. Thus v_{n-1} is adjacent to at least $n/2 - 1$ vertices between v_2 and v_{n-3}; the function $g_\ell(k)$ provides us with $n/2 - 1$ vertices that cannot be neighbors of v_1. Hence v_1 may have degree at most $n/2$, contradicting the assumption that the odd vertices all had higher degree. $\qquad \square$

Now, to complete the proof of Theorem 12, we take a graph G that satisfies the Chvátal condition; by Lemma 12.12, we may assume that n is even and $d_{n/2} \geq n/2$. Thus G has more than $n/2$ vertices of degree $n/2$ or higher, requiring that two such are consecutive in the Hamilton cycle. Then by Lemma 12.3, part 3, G is either pancyclic or bipartite. $\qquad \square$

We establish as a consequence the following:

Corollary 12.2. *Each of the following is sufficient to imply that G is pancyclic:*

1. *We have n odd, $1 \leq \delta \leq \frac{n+5}{6}$, and $e > \frac{1}{2}(n^2 - (2\delta + 1)n + 3\delta^2 + \delta)$;*
2. *We have n odd, $\frac{n+5}{6} \leq \delta \leq \frac{n-1}{2}$, and $e > \frac{1}{8}(3n^2 - 8n + 5) + \delta$;*
3. *We have n even, $1 \leq \delta \leq \frac{n+8}{6}$, and $e > \frac{1}{2}(n^2 - (2\delta + 1)n + 3\delta^2 + \delta)$;*
4. *We have n even, $\frac{n+8}{6} \leq \delta \leq \frac{n-1}{2}$, and $e > \frac{1}{8}(3n^2 - 10n + 16) + \delta$;*
5. *For any n, $\frac{n-1}{2} < \delta$, and $e \geq n^2/4$.*

Proof. We may assume that for some $k \leq (n-1)/2$ we have $d_k \leq k$ but $d_{n-k} < n-k$, as otherwise the conclusion follows from Theorem 12. Consider a graph G in which each vertex degree is as large as possible to be consistent with the existence of such a k. Then we have one vertex of degree δ, $k-1$ vertices of degree k (to ensure $d_k \leq k$), $n-2k$ vertices of degree $n-k-1$ to ensure that $d_{n-k} < n-k$, $k-\delta$ vertices of degree $n-2$, and no more than δ vertices of degree $n-1$ since any vertex of degree $n-1$ must be adjacent to the vertex of degree δ.

The idea is to show that even a graph with such a degree sequence does not have enough edges to meet the foregoing bounds, so that any graph that does meet those bounds will satisfy Chvátal's condition, and thus be pancyclic by the theorem. Let G be such a graph; we add these degrees and divide by 2.

For this graph, we have $e = (n^2 - (2k+1)n + 3k^2 - k + 2\delta)/2$. This value of e is quadratic in k, and we note that $\delta \leq k \leq \frac{n-1}{2}$. We first prove Item 2 of the corollary.

Setting up the inequality $e = (n^2 - (2k + 1)n + 3k^2 - k + 2\delta)/2 \geq \frac{1}{8}(3n^2 - 8n + 5) + \delta$ and solving for k, we find in fact that $k = \frac{n+5}{6} \leq k \leq \frac{n-1}{2}$. Using $\delta \leq k \leq \frac{n-1}{2}$, we get the result.

For $\delta \leq k \leq \frac{n+5}{6}$, we observe that the expression for e is largest when $k = \delta$, and making that substitution yields Item 1 of the corollary. Items 3 and 4 follow similarly, using $k \leq \frac{n}{2} - 1$. Finally, Item 5 is a direct consequence of the theorem. $\qquad \square$

That these bounds are best possible is shown in detail in [28]; we refer the interested reader to that paper for the constructions. This corollary also establishes Theorem 11.

A consequence of this corollary is the following less precise (but still useful) bound:

Corollary 12.3. *If G is a graph with n vertices and at least $\binom{n-1}{2} + 2$ edges, then G is pancyclic.*

Proof. Set $\delta = 1$ in Corollary 12.2. This yields (for n even or odd) $e > \frac{1}{2}(n^2 - 3n + 4)$. Careful factoring gives us

$$e > \frac{(n-1)(n-2) + 2}{2} = \binom{n-1}{2} + 1$$

as required. It is worth noting, as do Schmeichel and Hakimi [28], that a graph with this many edges must have $\delta \geq 2$ and has more edges than a complete bipartite graph on that number of vertices.

There is a conjecture due to Häggkvist [17] and independently to Mitchem and Schmeichel [24] to the effect that if G is a Hamiltonian graph on n vertices with minimum degree $\delta \geq (2n + 1)/5$, then G is pancyclic. This conjecture is proven for $n \geq 102$ in [1], and the interested reader is referred to that paper for more detail.

Now we prove that the condition of Fan in Theorem 5 implies that the graph is pancyclic, with only a few exceptions.

Theorem 13. *Suppose G is a 2-connected graph on n vertices with the property that any two vertices at distance 2 in the graph include at least one with degree at least n/2, then either G is pancyclic, or G is one of the three graphs $K_{n/2,n/2}$, $K_{n/2,n/2} - e$, or the graph F_n in Fig. 3.1.*

Proof. First we note that G contains a Hamilton cycle $(v_1, v_2, \ldots, v_n, v_1)$ by Theorem 5. By degree condition we have $\Delta(G) \geq n/2$. Without loss of generality, we may assume $d(v_1) \geq n/2$.

Now consider the vertices v_2 and v_n. If $v_2 v_n$ is an edge in G, then we have a cycle of length $n - 1$, specifically $(v_2, v_3, \ldots, v_n, v_2)$. Therefore G is pancyclic by Lemma 12.1. If $v_2 v_n$ is not an edge in G, then $D(v_2, v_n) = 2$, so $\max\{d(v_2), d(v_n)\} \geq n/2$. Without loss of generality, we may assume $d(v_n) \geq n/2$. If $d(v_1) \neq n/2$ and $d(v_n) \neq n/2$, then $d(v_1) + d(v_n) > n$. Hence, G is pancyclic by Lemma 12.2. If $d(v_1) = d(v_n) = n/2$, then by Lemma 12.3, G is pancyclic or bipartite unless conditions (1)–(4) in Lemma 12.3 all hold. In addition, it is straightforward to see that if G is bipartite, the degree condition together with the assumption $d(v_1) = d(v_n) = n/2$ implies that G is $K_{n/2,n/2}$ or $K_{n/2,n/2} - e$. In what follows we show that the degree condition together with conditions (1)–(4) imply that G is the graph F_n displayed in Fig. 3.1.

By Lemma 12.3 Part (1), G does not contain an $(n - 1)$-cycle. This implies that $v_j v_{j+2} \notin E(G)$ for any j. Since $D(v_j, v_{j+2}) = 2$, it follows that $\max\{d(v_j), d(v_{j+2})\} \geq n/2$. Therefore at least $n/2$ of the vertices of G have degree at least $n/2$. On the other hand, by Lemma 12.3 Part (3), at most $n/2$ vertices in G have degree at least $n/2$. So exactly $n/2$ vertices in G have degree at least $n/2$. By Lemmas 12.2 and 12.3

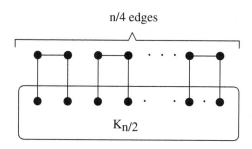

Fig. 3.1 Graph F_n

Part (3) we see that $n \equiv 0 \pmod 4$. Since $d(v_1) = d(v_n) = n/2$, it follows that $d(v_k) = n/2$ if $k \equiv 0, 1 \pmod 4$ otherwise $d(v_k) < n/2$.

Let

$$A = \{v_i \mid d(v_i) < n/2\} = \{v_i \mid i \equiv 2, 3 \pmod 4\}$$

and

$$B = \{v_i \mid d(v_i) = n/2\} = \{v_i \mid i \equiv 0, 1 \pmod 4\}.$$

Claim 1. If $v_i, v_j \in A$, then $v_i v_j \notin E(G)$ unless $j = i \pm 1$.

Suppose $v_i, v_j \in A$, and $v_i v_j \in E(G)$ but $j \neq i \pm 1$. Since either v_{i-1} or v_{i+1} is in A, without loss of generality, we may assume $v_{i+1} \in A$. If $v_{i+1} v_j \notin E(G)$, then $D(v_{i+1}, v_j) = 2$ which implies that $\max\{d(v_{i+1}), d(v_i)\} \geq n/2$, which contradicts $v_{i+1}, v_j \in A$. So $v_{i+1} v_j \in E(G)$. Now either v_{j-1} or v_{j+1} is in A. If $v_{j-1} \in A$, then $v_{j-2} v_{j+1}$ is an edge by Lemma 12.3 Part (4) and G contains the $(n-1)$-cycle $(v_i, v_j, v_{i+1}, v_{i+2}, \ldots, v_{j-2}, v_{j+1}, v_{j+2}, \ldots, v_i)$, a contradiction. The case $v_{j+1} \in A$ is similar.

Claim 2. If $v_i \in B$ and $v_j \in A$, then $v_i v_j \notin E(G)$ unless $j = i \pm 1$.

Suppose $v_i \in B$, $v_j \in A$, and $v_i v_j \in E(G)$ but $j = i \pm 1$. Since either v_{i-1} or v_{i+1} is in A, without loss of generality, we may assume $v_{i-1} \in A$. An argument similar to that described in Claim 1 shows that $v_{i-1} v_j$ is an edge in G. Now by Claim 1 we have $v_j = v_{i-2}$. This contradicts Lemma 12.3 Part (1).

By Claims 1 and 2 it is easy to see that the graph induced by B is $K_{n/2}$. Hence G must be the graph F_n displayed in Fig. 3.1. □

Theorem 14 ([2]). *Let G be a 2-connected graph on n vertices. Suppose that for every set of three independent vertices $x, y,$ and z, we have $d(x) + d(y) + d(z) \geq 3n/2 - 1$. Then G is pancyclic, $K_{n/2,n/2}$, $K_{n/2,n/2} - e$ or C_5.*

Proof. Note that G contains a Hamilton cycle $C_n = (v_1, v_2, \ldots, v_n, v_1)$ by Theorem 8. We first prove the following claim:

Claim. If $d(v_j)+d(v_{j+1}) \geq n$ for some j, then G is pancyclic, $K_{n/2,n/2}$ or $K_{n/2,n/2}-e$.

Proof of Claim. Note that if $d(v_j) + d(v_{j+1}) > n$, then G is pancyclic by Lemma 12.2. Now assume $d(v_j) + d(v_{j+1}) = n$. Without loss of generality we may assume $d(v_j) \leq n/2$. Then by Lemma 12.3, G is pancyclic or bipartite unless conditions (1)–(4) all hold for G. If conditions (1)–(4) hold for G, then by (2), $\{v_{j-2}, v_j, v_{j+2}\}$ is an independent set and by (3), $d(v_{j-2}), d(v_{j+2}) < n/2$. So $d(v_{j-2}) + d(v_j) + d(v_{j+2}) < 3n/2 - 3/2 < 3n/2 - 1$, a contradiction. Hence G is pancyclic or bipartite. If G is bipartite, from the degree condition it follows that G is $K_{n/2,n/2}$ or $K_{n/2,n/2} - e$. This completes the proof of claim.

We now assume G is not a 5-cycle and prove that $\Delta(G) \geq n/2$. If $\alpha(G) = 1$, then G is a complete graph and hence it is pancyclic. If $\alpha(G) \geq 3$, from the degree condition one can see that G is pancyclic. Now let $\alpha(G) = 2$. Then $\chi(G) \geq n/2$ because any color class contains at most 2 vertices. The assumptions $\alpha(G) = 2$ and G is not a 5-cycle imply that G is not an odd cycle. Therefore $\Delta(G) \geq \chi(G) \geq n/2$ by Theorem 3.

Now let $x \in V(G)$ with $d(x) = \Delta = \Delta(G) \geq n/2$. Let y and z be the vertices immediately preceding and succeeding x on the Hamilton cycle C_n. If $yz \in E(G)$, then G is pancyclic by Lemma 12.1. Hence, we assume $yz \notin E(G)$. If $d(x)+d(y) \geq n$ or $d(x) + d(z) \geq n$, then G is pancyclic, $K_{n/2,n/2}$ or $K_{n/2,n/2} - e$ by Claim 1. Now assume $d(y), d(z) \leq n - \Delta - 1$. Then $d(y) + d(z) \leq 2(n - \Delta) - 2 \leq n - 2$. So there exists a vertex $u \neq y, z$ such that $\{u, y, z\}$ is an independent set. Therefore $d(u) + d(y) + d(z) \geq 3n/2 - 1$. This leads to

$$d(u) \geq 3n/2 - 1 - (d(y) + d(z)) \geq 3n/2 - 1 - (2(n - \Delta) - 2) \geq \Delta + 1,$$

which is a contradiction.   □

3.3 Pancyclic Graph Products

It is natural to ask when a graph that arises in the course of study has a property such as pancyclicity. We have already discussed the question with regard to classes of Hamiltonian graphs; research has also been done on the existence of conditions that guarantee a product of graphs is pancyclic, for a few of the commonly defined graph products that we discussed in Sect. 1.3.

First, we consider the cartesian product, and in particular the prism. In [14], it is shown that:

Theorem 15. *If G is a 3-connected cubic graph on n vertices, then $G \times K_2$ has cycles of every even length from 4 up to $2n$; and if in addition G contains a 3-cycle, then $G \times K_2$ is pancyclic.*

This result has an elegant consequence. The *cube graph Q_n* is defined as the graph whose vertices are the binary strings of length n, with two vertices being adjacent

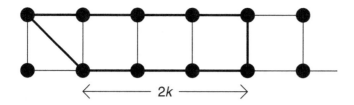

Fig. 3.2 Constructing the $2k + 1$-cycle

when the strings differ in precisely one place. We may also describe Q_n recursively; Q_1 is the complete graph on two vertices, and $Q_n = Q_{n-1} \times Q_1$.

Corollary 15.1. *For each $n \geq 2$, the graph Q_n contains cycles of every even length ℓ, $4 \leq \ell \leq 2^n$. Furthermore, we may add a single edge to Q_n to produce a pancyclic graph.*

Proof. It is known that Q_n is regular of degree n and has a Hamilton cycle (for $n \geq 2$). Thus, for $n \geq 3$, Q_n has a spanning subgraph G that is $C_{2^{n-1}} \times K_2$ generated by that Hamilton cycle. This subgraph G satisfies the conditions of Theorem 15, which proves the result for $n > 3$. It is simple to find 4-, 6-, and 8-cycles in Q_3, and Q_2 clearly has a 4-cycle.

Suppose we insert an edge in such a way to form a chord with any of the 4-cycles of Q_n. In particular, let S be a binary string of length $n - 2$, and consider the four vertices with strings $00 + S$, $01 + S$, $10 + S$, and $11 + S$. Here we use $+$ to denote concatenation. (This selection is without loss of generality, considering the abundance of automorphisms of Q_n.) Let us add the edge from $00 + S$ to $11 + S$.

Find a Hamilton path in Q_{n-1}, say $(v_0, v_1, v_{2^{n-1}-1})$. It is known that we can do this, and because Q_n is edge-transitive we may assume $v_0 = 0 + S$ and $v_1 = 1 + S$. This path gives us a ladder graph $P_{2^{n-1}} \times K_2$ in Q_n. The vertices of Q_n now may be written $(0, v_i)$ or $(1, v_i)$, for $0 \leq i \leq 2^{n-1} - 1$. The edge from $00 + S$ to $11 + S$ that we added earlier is in this notation the edge from $(0, v_0)$ to $(1, v_1)$. To construct our $2k + 1$-cycle, we start with the $2k$-cycle in Q_n that uses the second "rung" (edge from $(0, v_1)$ to $(1, v_1)$), the $k - 1$ edges joining $(0, v_i)$ to $(0, v_{i+1})$, $(1 \leq i \leq k - 1)$ the other $k - 1$ edges joining $(1, v_i)$ to $(1, v_{i+1})$, $(1 \leq i \leq k - 1)$, and the edge from $(0, v_k)$ to $(1, v_k)$. $\square$

The construction is illustrated in Fig. 3.2, below.

The authors of [27] consider strong products, and prove the following theorem:

Theorem 16. *If G is connected and has at least 2 vertices, then there is an m for which $G^{\boxtimes m}$ has a pancyclic ordering from each vertex.*

3.4 Open Problems

Many questions remain to be answered. It is known that the problem of determining whether a given graph is Hamiltonian is *NP*-complete, and so one would expect at least this difficulty in determining whether a given graph is pancyclic; however, there are classes of graphs for which the restriction of the problem to that class is polynomial time. Is there such a class of graphs for which determining the pancyclic property is still *NP*-complete? For what classes of graphs is the restriction of the pancyclic problem solvable in polynomial time?

Similar questions arise with random graphs (for a discussion of which, see [19]). What is the threshold value for the pancyclic property?

We also note that there exist other graph products besides those shown (e.g., the lexicographic product—see, among other sources, [18]); can one find better conditions on the factors that would imply that the product is pancyclic? Are there bounds on the exponent m in Theorem 16?

A graph with n vertices is (r)-*pancyclic* if it contains precisely r cycles of every length from 3 to n. In [38], Zamfirescu studies (2)-pancyclic graphs and finds the two (2)-pancyclic graphs of order 8, which are the smallest (2)-pancyclic graphs. In the same paper, Zamfirescu proves that there exist no (2)-pancyclic graphs on 9 or 10 vertices and presents a (2)-pancyclic graph of order n for each $n \in \{11, 13, 17, 19\}$. The existence of (r)-pancyclic graphs for $r \geq 3$ is still unknown.

Chapter 4
Minimal Pancyclicity

4.1 Introduction

As we pointed out in Chap. 3, for every positive integer $n \geq 3$ there is an integer $m(n)$ such that any pancyclic graph with n vertices must have at least $n + m(n)$ edges; a pancyclic graph with n vertices and $n + m(n)$ edges is called *minimal*. The difference $e(G) - v(G)$ is called the *excess* of the graph G, so $m(n)$ is the minimum excess for an n-vertex pancyclic graph. If a pancyclic graph contains exactly one cycle of every possible length, it is called *uniquely pancyclic*; we shall discuss such graphs in the following chapter.

The concept of minimal pancyclicity was introduced by Bondy, in the paper [3] where pancyclicity was first defined. He proposed the question, "What is the minimum number of edges in a pancyclic graph" with a given number of vertices. However, he did not use the phrase "minimally pancyclic" in that paper.

A popular conjecture is that one cannot decrease $m(n)$ by increasing n; in other words,

Conjecture 4.1. $m(n) \geq m(n-1)$.

Corollary 12.3 states that any graph with n vertices and more than $\binom{n-1}{2} + 1$ edges must be pancyclic, so we have an upper bound:

$$m(n) \leq \frac{n^2 - 3n}{2} + 3. \tag{4.1}$$

However, the minimum value is significantly smaller than this bound in most cases.

Suppose G is an $(n-1)$-vertex pancyclic graph with $(n-1) + m(n-1)$ edges. Add a new vertex y to G. Then select any edge (x, z) on the Hamilton cycle of G, and add edges (x, y) and (y, z) to G. The new graph has $(n-1) + m(n-1) + 2$ edges, and is pancyclic—all the old cycles still exist, and there is a Hamilton cycle formed by replacing (x, z) by (x, y) and (y, z) in the original. We have

© The Author(s) 2016
J.C. George et al., *Pancyclic and Bipancyclic Graphs*, SpringerBriefs in Mathematics, DOI 10.1007/978-3-319-31951-3_4

Theorem 17. $m(n) \leq m(n-1) + 1$.

If Conjecture 4.1 is true then $m(n)$ equals either $m(n-1)$ or $m(n-1) + 1$.

4.2   Minimal Pancyclic Graphs: Small Orders

The results in this section are taken from [13, 16]. We shall establish the value of $m(n)$ for $n \leq 37$.

Any pancyclic graph is Hamiltonian, so we shall represent an n-vertex pancyclic graph as a cycle

$$(a_1, a_2, \ldots, a_n, a_1)$$

together with some edges (which we shall call *chords*) of the form (a_i, a_j). The number of chords equals the excess of the graph.

We shall start by examining the possible patterns of cycles in small graphs, and determine the value of $m(n)$ for $n \leq 37$. We shall then establish upper bounds better than those of (4.1) for all cases.

Suppose we have a pancyclic graph on n vertices. It must contain at least $n - 2$ cycles—more, if some cycles are of equal length. We shall examine the possible patterns of cycles. The basic model is a cycle of length n with k chords, yielding $n + k$ edges in all. We shall work up through the values of k. The analysis is based on that in [23], with some modifications (the distinction we shall make below between types A and B is not drawn there).

In every case, we represent our graph as a circle (the Hamilton cycle) with the chords as straight lines. The segments of the outer circle may contain further vertices in addition to the vertices where the chords meet the Hamilton cycle; when it is necessary to specify the number of vertices in an arc, the easiest way is to write the number of *edges* in an arc next to the arc. The chords only have vertices at their ends, no intermediate vertices.

If the vertices of the Hamilton cycle are (in order) $a_1, a_2, \ldots, a_n$, then we say the chord $C = (a_i, a_j)$ bypasses the $j - i - 1$ vertices $\{a_{i+1}, a_{i+2}, \ldots, a_{j-1}\}$, which we shall call the *deficit* of the chord, and generates two cycles containing that one chord (and no other), namely $(a_i, a_{i+1}, \ldots, a_{j-1}, a_j, a_i)$ and $(a_j, a_{j+1}, \ldots, a_v, a_1, \ldots, a_i, a_j)$. When these two cycles are of different length, the shorter one is called the *small cycle* of C and the other is the *large cycle*. The number of vertices in the deficit is called the *deficiency* of the chord, so a chord of deficiency d generates cycles of lengths $d + 2$ and $n - d$. A chord with deficiency d is sometimes called a $(d + 2)$-chord, because it generates a $(d + 2)$-cycle. If $i < j$ we call a_i and a_j the *opening* and *closing* vertices of the chord (a_i, a_j).

Two properties of pairs of chords will occur later. We say chords *intersect* if each has one endpoint in the deficit of the other. In other words, if the Hamilton cycle is drawn as a circle and the chords are drawn as straight-line chords of the cycle, those

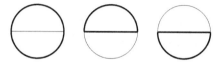

Fig. 4.1 Cycles in the case of one chord

two lines would intersect at an internal point. We say chord A *encloses* chord B if both endpoints of B are in the deficit of A, or if one is in the deficit and the other is also an endpoint of A.

4.2.1   Fewer Than Two Chords

If there are no chords, the graph contains only one (Hamilton) cycle, and the only pancyclic case is $n = 3$. If there is one chord, there are two further (one-chord) cycles. So, if the graph has only one chord, you obtain exactly three cycles, and if the graph is pancyclic it has $n \leq 5$. These cycles are illustrated in Fig. 4.1 (there are three drawings of the same graph, with the cycles shown in bold).

So $m(n) = 0$ if and only if $n = 3$, and $m(n) = 1$ if and only if $n = 4$ or 5.

4.2.2   Two Chords

If there are two chords, we distinguish three cases:

A. They do not cross, and have no common endpoint;
B. They have one common endpoint;
C. They cross.

Chords that cross are called *skew*. When necessary, we refer to two chords that are not skew as a nonskew pair.

In addition to the cycles that contain no chord or one chord, there may be new cycles that contain two chords. Let us count cycles in the three cases:

A. There is one new cycle, so together with the Hamilton cycle and the four one-chord cycles (two per chord) you have six cycles in total;
B. Again you get one new cycle, for six in total;
C. You get two new cycles, for seven in total.

All the new cycles are shown in Fig. 4.2.

This may allow us to find pancyclic graphs for up to $n = 9$ vertices, with only $n + 2$ edges, and in fact examples exist for $n \leq 8$ (see Fig. 4.5, below). However, there is no example for $n = 9$, as was shown by Shi [30]. So $m(n) = 2$ if and only if $6 \leq n \leq 8$.

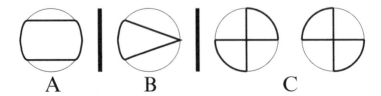

Fig. 4.2 Two chords

Fig. 4.3 The layout for nine vertices

A pancyclic graph on 9 vertices needs at least 7 cycles, so such a graph with two chords must be of type C. To see that no such graph is pancyclic, suppose the numbers of edges in the segments of the Hamilton cycle (*lengths* of the segments) are as shown in Fig. 4.3. Then $a+b+c+d = 9$. The lengths of the seven cycles are 9 (containing neither chord), $a+b+1, c+d+1$ (horizontal chord), $a+d+1, b+c+1$ (vertical chord), $a + c + 2, b + d + 2$ (both chords).

The only possible way to get a cycle of length 3 is if two adjacent segments have length 1, so we assume $a = b = 1$. Then $d = 9-a-b-c = 7-c$. The cycles (in the order listed above) have lengths $9, 3, 8, 9 - c, \ c + 2, c + 3, 10 - c$.

In order for there to be cycles of length 4, 5, 6, and 7, it is necessary that

$$\{9 - c, c + 2, c + 3, 10 - c\} = \{4, 5, 6, 7\}.$$

Without loss of generality, $c < d$. No case works (if $c = 1$, then $c+2 = 3$; if $c = 2$, then $10 - c = 8$; and if $c = 3$, then there is no 4-cycle). However, observe that the case $c = 2$ yields cycles of all length except 6, so it is easy to add one more chord and construct a 12-edge pancyclic graph on 9 vertices. Therefore $m(9) = 3$. It is also interesting to note that if we put $a = b = 1, c = 2$, and $d = n - 4$ we obtain a minimal pancyclic graph for $n = 6, 7, 8$.

4.2.3   Three Chords

If a graph has three chords, we classify by looking at the three pairs of chords. We refer to the configuration by the string of three letters corresponding to the three types of chord interaction. For example, type AAB is a graph in which two of the pairs of chords are type A (they do not cross, and have no common endpoint) and one pair is type B (they have a common endpoint). There are 14 types of graph: AAAi,

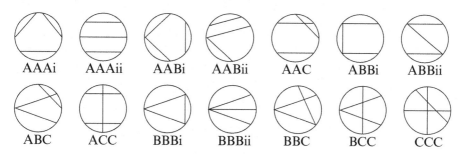

Fig. 4.4 Cases of three chords

AAAii, AABi, AABii, AAC, ABBi, ABBii, ABC, ACC, BBBi, BBBii, BBC, BCC, and CCC. (There are two types AAA, two types AAB, two types ABB (one where the three chords form a "C" pattern and one where they form a "Z"), and two types BBB (one where all three chords have a common endpoint and one where they form a triangle).) They are illustrated in Fig. 4.4.

The following table counts the number of cycles in a graph, in each of the 14 cases. $C(n)$ means the number of cycles involving exactly n chords.

	AAAi	AAAii	AABi	AABii	AAC	ABBi	ABBii
$C(0)$	1	1	1	1	1	1	1
$C(1)$	6	6	6	6	6	6	6
$C(2)$	3	3	3	3	4	3	3
$C(3)$	1	0	1	0	1	1	0
	11	10	11	10	12	11	10

	ABC	ACC	BBBi	BBBii	BBC	BCC	CCC
$C(0)$	1	1	1	1	1	1	1
$C(1)$	6	6	6	6	6	6	6
$C(2)$	4	5	3	3	4	5	6
$C(3)$	1	2	1	0	1	1	2
	12	14	11	10	12	13	15

From the table, it is clear that one can try to construct pancyclic graphs on n vertices and $n + 3$ edges for $10 \leq n \leq 17$. The cases of $n = 10, 11, 12$ were constructed in [32], and 14 was given in [23].

Using the same technique as we did for nine vertices, we can find an example for $n = 13$, and essentially the same method works for all cases with $10 \leq n \leq 14$. The diagram is and the graph has cycles of lengths 3, 4, 5, 6, 7, 8, $n - 5$, $n - 4$, $n - 3$, $n - 2$, $n - 1$ and n.

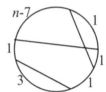

For $n > 14$, only types BCC, ACC, and CCC need be discussed; in fact, type CCC can be eliminated immediately, as no cycle involves fewer than either two chords and two segments or one chord and three segments, so a cycle of length 3 is impossible.

BCC has 13 cycles, so $n = 14$ or 15 could be possible, but the interesting case ($n = 15$) would be uniquely pancyclic, and that is ruled out in [23].

ACC has 14 cycles. We would need 13 different lengths for case $n = 15$, and all 14 different for $n = 16$. In the following diagram, lower-case letters represent the number of edges in a segment, while upper-case letters represent endpoints. In order to represent cycles, we list the upper-case letters in order, and assume that the other vertices lying in the segment between two consecutive upper-case vertices are included. For example, if $c = 1, d = 2$, and vertex A lies in the segment XY, then WXYW would represent the 4-cycle (W,X,A,Y,W).

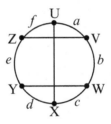

Without loss of generality we can assume $a = f = 1$, in order to get a cycle of length 3. The only possible cycles of length 4 are WXYW (which would imply $c + d = 3$, without loss of generality $c = 2, d = 1$), UVWXU ($b = c = 1$), VWYZV ($b = e = 1$), or UXYZU (mirror image of UVWXU). So we can assume either $b = 1$, and one of $c = 1$ or $e = 1$, or $c = 2$ and $d = 1$.

Assume $b = 1$. If $c = 1$, then VWXYZV and UXYZVU are both length $3+d+e$, while UVWYZU and UXWYZU are both length $4 + e$; in both cases we have at most 12 different lengths. If $e = 1$, then UVWYXU and UXYZVU are both length $4 + d$, and UXWVZU and UXWYZU are both length $4 + c$; again we have at most 12 different lengths.

Finally, set $c = 2, d = 1$. Then UVWYZU and UXYZVU are both length $4 + e$, and UXYZU and VWYZV are both length $3 + e$; there are at most 12 different lengths.

Therefore cases $n = 15$ and $n = 16$ both require at least four chords. Suitable constructions are found in [32], but see also the following section.

Figure 4.5 shows examples of minimal pancyclic graphs up to 14 vertices.

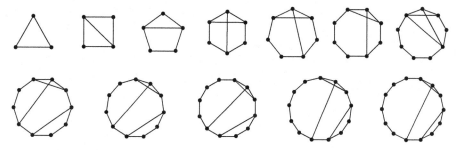

Fig. 4.5 Minimal pancyclic graphs up to order 14

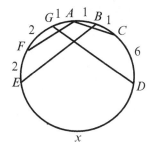

Fig. 4.6 Graph used to construct examples of orders 15–24

4.3 Four Chords

There are a large number of possible configurations for four chords. Several of them have been examined. Figure 4.6 shows one of the configurations. It represents a graph with $x + 13$ vertices, all on the outer circle. Some vertices are named, while the numbers show the number of edges on the segment between two labeled vertices: for example, there are six edges, and consequently five unlabeled vertices, between C and D.

The graph contains cycles of the following lengths:

Length	Cycle		Length	Cycle
3	ABCA		n	ABCDEFGA
4	AFGA		$n-1$	ACDEFGA
5	ABEFA		$n-2$	ABCDEFA
6	ACBEFA		$n-3$	ACDEFA
7	ABEFGA		$n-4$	ABEDCA
8	ACBEFGA		$n-5$	BCDEB
9	ACDGA		$n-6$	ACBEDGFA
10	ABCDGA		$n-7$	ABEDGFA
11	ACDGFA		$n-8$	AFEDGA
12	ABCDGFA		$n-9$	ABEDGA
13	BCDGEB		14	ABEGDCA

It will be observed that all lengths from 3 to n inclusive are represented at least once, provided $1 \leq x \leq 10$. So a minimal pancyclic graph has $n + 4$ edges (that is, $m(n) = 4$) when $15 \leq n \leq 24$.

It is reported in [16] that an exhaustive search was conducted that shows there is no pancyclic graph on 25 or more vertices with four or fewer chords. So we have established the value of $m(n)$ for all $n \leq 24$.

4.4 Five Chords

There are a large number of possible configurations for five chords. Figure 4.7 shows one of the configurations. It represents a graph with $n = x + 21$ vertices. The graph contains cycles of the following lengths:

Length	Cycle		Length	Cycle
3	DEFD		n	ABCDEFGHIJA
4	ABCA		$n - 1$	ABCDFGHIJA
5	ACDEJA		$n - 2$	ACDEFGHIJA
6	ACDFEJA		$n - 3$	ACDFGHIJA
7	ABCDEJA		$n - 4$	DEJIHGFD
8	ABCDFEJA		$n - 5$	EFGHIJE
9	ABHIJA		$n - 6$	ABHGFEJA
10	BCDEJIHB		$n - 7$	BCDEFGHB
11	BCDFEJHB		$n - 8$	BCDFGHB
12	ABHIJEDCA		$n - 9$	ABHIGFDEJA
13	ABHIJEFDCA		$n - 10$	ABHIGFEJA
14	GHIG		$n - 11$	BCDEFGIHB
15	ABHGIJA		$n - 12$	ABCDEFGIJA
16	BCDEJIGHB		$n - 13$	ABCDFGIJA
17	BCDFEJIGHB		$n - 14$	ACDEFGIJA
18	ABHGIJEDCA		$n - 15$	ACDFGIJA
19	ABHGIJEFDCA		$n - 16$	DEJIGFD
			$n - 17$	EFGIJE

This proves that $m(n) = 5$ for $25 \leq n \leq 37$.

4.5 More General Bounds for Pancyclics

In 1978, Sridharan [32] claimed to find the number of edges in a minimal pancyclic graph for all orders. The paper purports to give constructions of pancyclic graphs of all orders. However, a number of orders are omitted, and it appears that the author

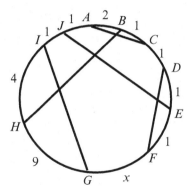

Fig. 4.7 Graph used to construct examples of orders 25–37

assumed Conjecture 4.1 to fill in the gaps. Moreover, it is never proven that the graphs constructed are minimal. The examples given are good, but not perfect; for example, it is stated that a pancyclic graph with 13 or 14 vertices requires at least four chords, while we know that the minimum is 3; a similar problem occurs for orders 21–24.

In his review of Sridharan's paper in Mathematical Reviews, Bondy said "Implicit in the claim (that the paper settles the problem suggested by Bondy) is the assumption that a minimal pancyclic graph is a pancyclic graph with as few edges as possible. This is easily seen to be false." Assuming that Bondy meant "as few edges as possible, given the number of vertices," this is certainly what Sridharan assumes, and is what the phrase is usually taken to mean. It is also exactly what Bondy's question asked for, although he did not use the phrase "minimal pancyclic" in his paper.

Sridharan's construction involves graphs with chords of deficiency $1, 2, 4, 8, \ldots$, that is to say $2^0, 2^1, 2^2, 2^3, \ldots$, no two of which intersect and no one encloses any other. Define S_n^k to be a graph consisting of a Hamilton cycle $(a_1, a_2, \ldots, a_{n-1}, a_n, a_1)$, together with chords $A_0, A_1, \ldots, A_k$ of the kind described.

Since the set of deficiencies of the chords in S_n^k is $D = \{1, 2, 2^2, \ldots, 2^k\}$, S_n^k contains cycles of lengths $2^i + 2$, $0 \le i \le k$, and also of all lengths $n - j$, where $j \in D$, that is all orders from $n - (2^{k+1} - 1)$ to $n - 1$, as well as the Hamilton cycle. This is enough to make S_n^k pancyclic provided $n - (2^{k+1} - 1) \le 5$; otherwise there is no cycle of length 5. Similarly, 7, 8, and other orders not of the form $2^i + 2$ will be missing in later cases. Sridharan [32] approaches this problem by adding some more chords later in the process.

Sridharan's proof starts as follows.

Start with a Hamilton cycle $(a_1, a_2, \ldots, a_n, a_1)$, which is itself a pancyclic graph when $n = 3$. Add a chord $A_0 = (a_1, a_3)$, with deficiency 1. The resulting graph is pancyclic when $n = 4, 5$. Now add a chord with deficiency 2, namely $A_1 = (a_{n-2}, a_1)$. This provides a pancyclic graph for $n = 6, 7, 8$.

The next case adds a chord $A_2 = (a_{n-7}, a_{n-2})$ of deficiency $2^2 = 4$. Chords A_0 and A_2 will intersect in the case $n = 9$, but the resulting graph is pancyclic.

This example also works for $n = 10, 11, 12$. The resulting graph is not pancyclic when $n = 13$ or 14, because there is no cycle of length 5; we saw pancyclic graphs of excess 3 for those orders in Sect. 4.2.3 above, using a different construction.

For cases $13 \leq n \leq 20$, $A_3 = (a_3, a_{12})$ (with deficit 2^3) is added. Now there are some problems. When $v = 13$, chords A_1 and A_3 intersect. When $n = 13$ or 14, A_3 encloses A_2. And when $15 \leq n \leq 18$, chords A_2 and A_3 intersect. This means that, even when $n - (2^{k+1} - 1) \leq 5$, we cannot construct a cycle of length 5 by deleting the deficits of the chords. We carried out a computer analysis of Sridharan's constructions for $13 \leq n \leq 18$. In all cases except $n = 17$, the resulting graph is pancyclic. When $n = 17$, the cycles in the graph are as follows (in each case, the cycles listed contain no chord other than those listed):

containing A_0, (a_1, a_2, a_3, a_1) and $(a_1, a_3, a_4, a_5 \ldots, a_{17}, a_1)$, lengths 3 and 16;
containing A_1, $(a_1, a_{15}, a_{16}, a_{17}, a_1)$ and $(a_1, a_2, a_3, \ldots, a_{15}, a_1)$, lengths 4 and 15;
containing A_2, $(a_{10}, a_{11}, a_{12}, a_{13}, a_{14}, a_{15}, a_{10})$
and $(a_1, a_2, a_3, \ldots, a_{10}, a_{15}, a_{16}, a_{17}, a_1)$, lengths 6 and 13;
containing A_3, $(a_1, a_2, a_3, a_{12}, a_{13}, a_{14}, a_{15}, a_{16}, a_{17}, a_1)$ and $(a_3, a_4, a_5, \ldots, a_{12}, a_3)$, lengths 9 and 10;
containing A_0 and A_1, $(a_1, a_3, a_4, a_5 \ldots, a_{15}, a_1)$, length 14;
containing A_0 and A_2, $(a_1, a_3, a_4, a_5 \ldots, a_{10}, a_{15}, a_{16}, a_{17}, a_1)$, length 12;
containing A_0 and A_3, $(a_1, a_3, a_{12}, a_{13}, a_{14}, a_{15}, a_{16}, a_{17}, a_1)$, length 8;
containing A_1 and A_2, $(a_1, a_2, a_3, a_4, a_5 \ldots, a_{10}, a_{15}, a_1)$, length 11;
containing A_1 and A_3, $(a_1, a_2, a_3, a_{12}, a_{13}, a_{14}, a_{15}, a_1)$, length 7;
containing A_2 and A_3, $(a_1, a_2, a_3, a_{12}, a_{11}, a_{10}, a_{15}, a_{16}, a_{17}, a_1)$
and $(a_3, a_4, a_5 \ldots, a_{10}, a_{15}, a_{14}, a_{13}, a_{12}, a_3)$, lengths 9 and 12;
containing A_0, A_1 and A_2, $(a_1, a_3, a_4, a_5 \ldots, a_{10}, a_{15}, a_1)$, length 10;
containing A_0, A_1 and A_3, $(a_1, a_3, a_{12}, a_{13}, a_{14}, a_{15}, a_1)$, length 6;
containing A_0, A_2 and A_3, $(a_1, a_3, a_{12}, a_{11}, a_{10}, a_{15}, a_{16}, a_{17}, a_1)$, length 8;
containing A_1, A_2 and A_3, $(a_1, a_2, a_3, a_{12}, a_{11}, a_{10}, a_{15}a_1)$, length 7;
containing A_0, A_1, A_2 and A_3, $(a_1, a_3, a_{12}, a_{11}, a_{10}, a_{15}, a_1)$, length 6.
There is no cycle of length 5.

In cases $n = 19, 20$, the argument in [32] does prove that the resulting graph is pancyclic, because in those cases the chords do not intersect.

In the cases starting from 21, Sridharan observes that the cycle of length 5 is a problem, as will be 7, 8, and other small orders not of the form $2^i + 2$, so a chord $A_4^* = (a_{12}, a_{v-7})$ is introduced. This chord provides cycles of length $n - 18$ and 20. The chords A_0, A_1, A_2, A_3 are all enclosed in the cycle of length 20, so these chords together with A_4^* provide cycles of all lengths formed by subtracting a sum of their deficits from 20, that is all lengths from 5 (that is, 20–15) to 19. Using the deficits of A_0, A_1, A_2, and A_3 we can form cycles of all lengths from $n - 1$ to $n - 15$, so all needed cycles are provided for orders 21 through 36, with an excess of 5.

The next step is to add the next chord, $A_5 = (12, 29)$. That is , it shares a vertex with A_4, and has deficit 2^4. It is easy to see that the chords used so far provide pancyclic graphs of orders 37–52 with excess 6.

Now the process breaks down. The next chord, A_6 of deficit 2^5, would be (a_{29}, a_{62}). If $n < 69$, the chord will either intersect or enclose some of the earlier chords—and, if $n < 62$, we would have to assume that the second vertex of A_6 is reduced modulo n. But the proof of pancyclicity assumes that none of the chords of deficit $2^0, 2^1, 2^2, 2^3, \ldots$ enclose or intersect. And some of the resulting graphs are not pancyclic. Again, we carried out a computer analysis for $53 \leq n \leq 68$. The only pancyclic cases are those with 53, 60, and 65 vertices. The missing cycle lengths in the other cases are as follows:

Order	Missing cycles		Order	Missing cycles
54	21		55	22
56	21, 23		57	22
58	21, 22		59	21
61	22, 24, 26		62	25, 27
63	24, 26, 28		64	25, 26, 27,28
66	21, 29, 30		67	22, 28, 29, 30, 31
68	21, 22, 29, 30			

We shall present a slight modification of Sridharan's proof. We consider a graph consisting of a cycle $H_n = (a_1, a_2, \ldots, a_n, a_1)$ together with some chords. We define a sequence of vertices x_k, where $x_0 = a_1$ and $x_k = a_{2^k+k}$. We now define the chord $A_k = (x_k, x_{k+1})$, provided $n > 2^{k+1}+k$. Then A_k has deficiency 2^k, and the chords do not intersect but each chord opens with the previous chord's closing vertex. Finally, we define A_k^* to be the chord (x_k, x_0). A_k^* has deficiency $n - 2^k - k$.

If $2^k + k \leq n$, none of the chords $A_0, A_1, \ldots, A_{k-1}$ will overlap, so we can add all those chords to H_n, and the resulting graph has cycles of all lengths $n - d$, where $1 \leq d \leq 2^k - 1$ (d ranges through the sums of deficiencies of the chords), as well as n and the orders $\{2^i + 2 : 0 \leq i \leq k - 1\}$.

We now add A_k^* to the graph. This introduces cycles of length $n - 2^k - k + 2$ and $2^k + k$. Provided $k > 1$, the cycle of length $2^k + k$ encloses each of $A_0, A_1, \ldots, A_{k-1}$, so we also get cycles of length $2^k + k - d$, where $1 \leq d \leq 2^k - 1$; in other words, we get all lengths from $k + 1$ to $2^k + k - 1$. (The case $k = 1$ must be omitted because A_1^* is identical to A_0.)

Let us define

$$G_k = H_n \cup A_0 \cup A_1 \cup \ldots \cup A_{k-1} \cup A_k^*$$

for all orders n from $2^k + k + 1$ to $2^{k+1} + k$, and

$$G_k = H_n \cup A_0 \cup A_1 \cup \ldots \cup A_{k-1} \cup A_k$$

when $n \geq 2^{k+1} + k + 1$. Then, from $k = 2$ onward, G_k contains cycles of every length from $k + 1$ to n, as well as lengths $2^i + 2$ for $0 \leq i < k$. It follows that G_k is pancyclic for $2 \leq k \leq 4$, for the given values of n, and it is easy to see that G_0 is pancyclic for $n = 4, 5$ and G_1 is pancyclic for $n = 6, 7, 8$. We have pancyclic graphs with the following parameters:

Vertices	Graph	Edges	Excess
$n = 3$	K_3	3	0
$4 \leq n \leq 5$	G_0	$n + 1$	1
$6 \leq n \leq 8$	G_1	$n + 2$	2
$9 \leq n \leq 12$	G_2	$n + 3$	3
$13 \leq n \leq 20$	G_3	$n + 4$	4
$21 \leq n \leq 36$	G_4	$n + 5$	5

However, G_4 is not pancyclic in the largest case, $n = 37$, because it contains no cycle of length 5.

When $n \geq 37$, the chords A_4 and A_4^* are different, so one can include both of them in one graph. Consider the graph formed by adding A_0, A_1, A_2, A_3, A_4, and A_4^* to H_n. Chords A_0, A_1, A_2, A_3, and A_4 provide cycles of lengths 3, 4, 6, 10, 18, and $n - d : 1 \leq d \leq 2^5 - 1 = 31$; H_n provides length n; and A_0, A_1, A_2, A_3, and A_4^* provide orders from 5 to $2^4 + 4 = 20$. So the resulting graph is pancyclic provided $n - 31 \leq 21$, that is $n \leq 52$. So we have a pancyclic graph of excess 6 when $37 \leq n \leq 52$.

The graph G_j has cycles of all lengths from $n - (2^{j+1} - 1)$ upwards, and the addition of A_4^* provides cycles of all lengths up to 20. So, for $5 \leq j \leq 20$, $G_j \cup A_4^*$ is a pancyclic graph with an excess of $j + 2$ provided $n - (2^{j+1} - 1) \leq 21$, that is $n \leq 2^{j+1} + 20$, and provided none of $A_0, A_1, \ldots, A_j$ overlap. So we have a pancyclic graph of excess $j + 2$ when $2^j + j + 1 \leq n \leq 2^{j+1} + 20$. Note that the "+20" addition to the upper limit of n means that a few of the smallest values for which we just established an excess of $j + 2$ will in fact be covered by the construction for excess $j + 1$, so we actually have

Theorem 18. *When $j \leq 21$, there is a pancyclic graph on n vertices with $n + j + 2$ edges whenever $2^j + 21 \leq n \leq 2^{j+1} + 20$.*

So we have a bound for all orders up to $n = 2^{22} + 20 = 4{,}194{,}234$. Here are some of the smaller cases; we have found pancyclic graphs with the following parameters:

Vertices	Edges	Excess
$37 \leq n \leq 52$	$n + 6$	6
$53 \leq n \leq 84$	$n + 7$	7
$85 \leq n \leq 148$	$n + 8$	8
$149 \leq n \leq 276$	$n + 9$	9
$277 \leq n \leq 532$	$n + 10$	10
$533 \leq n \leq 1044$	$n + 11$	11
$1045 \leq n \leq 2068$	$n + 12$	12

When $j = 22$, there will be no cycle of length 21. However, the addition of the chord A_5^* provides cycles of all lengths from 6 to 37. Note that when chord A_j^* is added, it and chords $A_0, A_1, \ldots, A_{j-1}$ together form cycles of all lengths from $j + 1$ to $2^j + j$. In general, to avoid missing a cycle of length $2^h + h + 1$, chord A_{h+1}^* is needed when $n > 2^j + j + 1$, where $j = 2^h + h + 1$. This means that when $2^{(2^h + h + 1)} + 2^h + h + 2 \leq n \leq 2^{(2^h + h + 2)} + 2^h + h + 1$, we add to H_n the chords $A_0, A_1, \ldots, A_{(2^h + h + 1)}$, and $A_4^*, A_5^*, \ldots, A_{h+1}^*$ to achieve a pancyclic graph.

So we have

Theorem 19. *When*

$$2^{(2^h + h + 1)} + 2^h + h + 2 \leq n \leq 2^{(2^h + h + 2)} + 2^h + h + 1,$$

there is a pancyclic graph on n vertices with $n + 2^h + 2h$ edges. So the excess for a minimal pancyclic graph with n as stated is at most $2^h + 2h$.

Chapter 5
Uniquely Pancyclic Graphs

5.1 Introduction

Recall that a pancyclic graph is called *uniquely pancyclic*, or UPC, if it contains *exactly* one cycle of every possible length. In 1973, Roger Entringer asked (see [3], p. 247), for what orders do UPC graphs exist? The question remains unsolved, and it is not even known whether the number of UPC graphs is infinite.

A uniquely pancyclic graph with n vertices must contain precisely $n - 2$ cycles, one of each length $3, 4, \ldots, n$. One of these will be the Hamilton cycle, and the others must involve chords in that cycle. As we have seen before, each chord generates two more cycles; combinations of two chords generate one or two more cycles, and sets of three or more chords may generate more cycles, depending on their layout. We shall examine the smallest cases individually.

It is clear that an n-vertex pancyclic graph with no chords can contain only one cycle, namely the Hamilton cycle, and $n = 3$. So K_3 is trivially a UPC graph.

5.2 Small Cases

In this section we find all UPC graphs with two or fewer chords. These results were originally discovered by Shi [30].

Say a pancyclic graph contains one chord. Then it contains three cycles: the Hamilton cycle, and the two cycles including the chord. So the graph must contain at most five vertices, and to be UPC, exactly five. Again there is obviously only one example, with five vertices and a chord of deficiency 1.

Suppose there are two chords. The possibilities are illustrated in Fig. 5.1; as usual, the letters represent the number of vertices between the chord endpoints. Each chord will generate two one-chord cycles, and the two chords will lie together

© The Author(s) 2016
J.C. George et al., *Pancyclic and Bipancyclic Graphs*, SpringerBriefs
in Mathematics, DOI 10.1007/978-3-319-31951-3_5

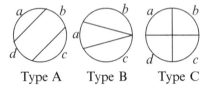

Fig. 5.1 Possibilities for two chords

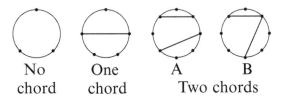

Fig. 5.2 UPC graphs with two or fewer chords

in either one cycle (if they are disjoint or have a common endpoint, that is they are type A or B) or two cycles if they are skew (type C). In the first case we must have $n = 8$; in the second, $n = 9$.

In case A, $a + b + c + d = 4$ and neither a nor c can be zero. The cycle lengths are $a + 2, b + c + d + 4, c + 2, a + b + d + 4, b + d + 4$, and 8. As there must be one cycle of length 3, either a or c is 1, but not both; let us choose a. Then

$$\{b + c + d + 4, c + 2, b + d + 5, b + d + 4\} = \{4, 5, 6, 7\}.$$

Now $b + c + d = 3$, and $c > 1$, so the possibilities are $c = 2, b = 1, d = 0$, $c = 2, b = 0, d = 1$ or $c = 3, b = 0, d = 0$. The first two are isomorphic, and the third case gives two cycles of length 5, so there is one example of type A, shown in Fig. 5.2 and labeled A.

For case B, we must have $a + b + c = 5$ and neither b nor c can be 0. The cycle lengths are $a + 3, b + 2, c + 2, a + c + 3, a + b + 3$, and 8. To get a 3-cycle we need $a = 0, b = 1$, or $c = 1$, and only one of these can be true. Without loss of generality assume $b < c$. The possibilities are $a = 0, b = 2, c = 3$, which gives two 5-cycles, $a = 1, b = 1, c = 3$, which also gives two 5-cycles, and $a = 2, b = 1, c = 2$, which works. The solution is shown in Fig. 5.2 and labeled B.

Finally, in case C, there are 7 cycles, so $n = 9$ and $a + b + c + d = 5$. The only way to get a 3-cycle is for two adjacent letters on the cycle each to have value zero. Let us assume $a = b = 0$. But then the 7 cycle lengths are 3, 8 (including the horizontal chord), $c + 3, d + 3$ (with the vertical chord), $c + 4, d + 4$ (with both chords), and 9. But c and d must add to 5; if $\{c, d\} = \{0, 5\}$ there are two cycles of length 9; if $\{c, d\} = \{1, 4\}$ there are two cycles of length 8; and if $\{c, d\} = \{2, 3\}$ there are two cycles of length 6. So there is no type 3 example.

5.3 Outerplanar UPC Graphs

A Hamiltonian graph is called *outerplanar* if it can be drawn with a Hamilton cycle on the outside with all other edges inside the cycle, without any crossings. Clearly all outerplanar graphs are planar, but the converse is not true: K_4 is an obvious counterexample. All four UPC graphs we found in the previous section are outerplanar; Shi [30] showed that these are the only such examples; we shall prove this result below.

Suppose the graph G consists of a Hamilton cycle $(x_1, x_2, \ldots, x_n, x_1)$ and some chords.

Consider two chords (x_a, x_b) and (x_c, x_d). If $a < c < d < b$, then either (x_c, x_d) lies in the small cycle or the large cycle of (x_a, x_b). If it lies in the small cycle, we say (x_a, x_b) *encloses* (x_c, x_d); if (x_c, x_d) lies in the large cycle, we say (x_c, x_d) *is exterior to* (x_a, x_b). A chord is *strict* if it encloses no other chord and is not skew to any other chord. Obviously, if there is a chord with a small cycle of length t, the long cycle will have length $n - t + 2$, so $n - t + 2 > t$, or $n > 2t - 2$. Putting it another way, if there is a chord of deficiency k, then $v > 2k + 2$. So we have the following lemma:

Lemma 13.1. *If G is a UPC graph containing a chord of deficiency k, then $n > 2k + 2$.*

If all chords are strict, an outerplanar graph with m chords will contain $2^m + m$ cycles—two generated by each chord, and one for every other combination of chords. If some chords are enclosed, and $m > 2$, the number of cycles will be smaller: if C_1 encloses C_2, there will be no cycle containing both C_1 and C_2 together with any other chord. Skew chords may produce more cycles, but of course they do not exist in outerplanar graphs.

Lemma 13.2 ([31]). *If an outerplanar UPC graph G contains m chords, and $m \neq 2$, then it contains a strict chord of deficiency 2^i for each i, $0 \leq i \leq m - 1$.*

Proof. A pancyclic graph on n vertices contains one Hamilton cycle and one cycle of length $n - 1$. Suppose $(x_1, x_2, \ldots, x_n)$ is the Hamilton cycle and x_i is the lone vertex not contained in the $(n - 1)$-cycle. If no chords are to cross, the only possible $(n - 1)$-cycle is $(x_1, x_2, \ldots, x_{i-1}, x_{i+1}, \ldots, x_n, x_1)$, so there is a chord of deficiency 1, namely (x_{i-1}, x_{i+1}). Similarly, we see that there must be a chord of deficiency 2. So the lemma is true for $i = 0, 1$.

We proceed by induction. Suppose the outerplanar UPC graph has Hamilton cycle $(x_1, x_2, \ldots, x_n, x_1)$ and m chords, $m \geq 3$; and assume it contains a chord C_i of deficiency 2^i for $0 \leq i \leq k - 1$. By using a combination of those chords, we can construct cycles of every length $n - S$, where S is any sum of those deficiencies. By adding different selections from $\{2^0, 2^1, \ldots, 2^{k-1}\}$, we can make all the integers from 1 to $2^k - 1$, so we have cycles of every length from n down to $n - 2^k + 1$. These cycles include all cycles that contain one or more of $C_1, C_2, \ldots, C_{k-1}$.

Now consider the cycle in G, K say, of length $n - 2^k$, and select the chord C that is in K and has the largest deficiency of all chords in K. Clearly C is not one of

$\{C_1, C_2, \ldots, C_{k-1}\}$. Moreover K contains no other chord than C; for, if it contained another chord (x_i, x_j), the cycle obtained from K by replacing that chord by the edges $(x_i, x_{i+1}), \ldots, (x_{j-1}, x_j)$ would be longer than $n - 2^k$, so it would duplicate a length that we already have. So K is one of the two cycles generated by C. So C has deficiency 2^k; in other words, we could denote C by C_k, provided it is strict.

It remains to prove that C is strict. Suppose not; it must enclose some other chords with smaller deficiencies; the candidates are $\{C_1, C_2, \ldots, C_{k-1}\}$. Either C_{k-1} is enclosed or it is not.

First, suppose C_{k-1} is enclosed. Let j be the smallest number such that each of $C_j, C_{j+1}, \ldots, C_{k-1}$ is enclosed by C. The sum of the deficiencies of $C_j, C_{j+1}, \ldots, C_{k-1}$ is

$$\sum_{i=j}^{k-1} 2^i = 2^j (2^{k-j} - 1),$$

and the small cycle of C has length $2^k + 2$, so the cycle obtained from the small cycle of C by omitting the deficits of $C_j, C_{j+1}, \ldots, C_{k-1}$ has length

$$2^k + 2 - 2^j(2^{k-j} - 1) = 2^k + 2 - (2^k - 2^j) = 2^j + 2,$$

which is also the length of the small cycle of C_j. So G is not UPC.

So we assume C_{k-1} is not enclosed. Suppose C_j is the shortest chord enclosed by C, and suppose C encloses $C_j, C_{j+1}, \ldots, C_{j+r-1}$, but not C_{j+r}, for some j and r, where $1 \le r \le k-j$. By assumption, C does not enclose the chords $C_0, C_1, \ldots, C_{j-1}$. Using different combinations of the chords $\{C_0, C_1, \ldots, C_{j-1}, C\}$ we can find the unique $(n - i)$-cycle in G for each $0 \le i \le 2^k + 2^j - 1$. Hence G has no other chords with deficiencies less than $2^k + 2^j$. This implies that the $(n - 2^k - 2^j)$-cycle in G, say K^*, has a chord not in $\{C_0, C_1, \ldots, C_{k-1}, C\}$. For otherwise K^* is a small cycle; by Lemma 13.1, $n > 2(2^k) + 2 = 2^{k+1} + 2$, whence $n - 2^k - 2^j = 2^k + 2$. Now the cycle containing only the chords C_k and C_{j+r} and the cycle containing only the chords $C_k, C_j, C_{j+1}, \ldots, C_{j+r-1}$ are both of length $n - 2^k - 2^{j+r}$, which is a contradiction.

A discussion similar to that of the $(n - 2^k)$-cycle yields that G contains exactly one chord of deficiency $2^k + 2^j$. We denote this chord by C^*. Now we consider two cases.

If C_j is enclosed by C^*, then C is also enclosed by C^*. It follows that the cycle containing only C and the cycle containing only C_j and C^* are both of length $2^k + 2$, a contradiction.

If C_j is not enclosed by C^*, then C is not enclosed by C^*. It follows that the cycle containing only C and C_{j+r} and the cycle containing only $C^*, C_j, C_{j+1}, \ldots, C_{j+r-1}$ are both of length $n - 2^k - 2^{j+r}$, a contradiction.

Therefore the chord C is strict. Thus G contains exactly one chord of deficiency 2^k. Now the lemma follows by the principle of induction. $\square$

Theorem 13 ([31]). *There are no outerplanar UPC graphs with more than two chords.*

Proof. Suppose G is an outerplanar UPC graph with m chords, $m \geq 3$. From Lemma 13.2, all chords are strict, so there are $2^m + m$ cycles. Therefore a UPC graph with m chords will have $2^m + m + 2$ vertices. The set of deficiencies of the chords is

$$D = \{2^0, 2^1, \ldots, 2^{m-1}\}.$$

The graph will contain cycles of length $d + 2$, for $d \in D$, formed from a chord and its deficit. These cycles will therefore have lengths $3, 4, 6, \ldots$; the length 5 is not represented. There will be also cycles of all lengths $v - s$, where s is a sum of some members of D, with each member occurring at most once in the sum. The shortest length among these cycles is

$$n - \sum_{i=0}^{m-1} 2^i = (2^m + m + 2) - (2^m - 1) = m + 3.$$

As $m \geq 3$, the length 5 does not occur among these lengths either. So none of these graphs is pancyclic. $\qquad\square$

5.4 More General UPC Graphs

Lemma 14.1. *Suppose G is a UPC graph and C is a chord of deficiency 2 in G. Then C is not skew to the other chords in G.*

Proof. Let $C = (x_1, x_4)$ and suppose that there was a chord skew to C, say (x_2, x_i). Then the cycles $(x_2, x_3, \ldots, x_i, x_2)$ and $(x_2, x_1, x_4, \ldots, x_i, x_2)$ both have length $i - 1$, contradicting the uniqueness property. A chord (x_3, x_j) is similarly impossible. $\quad\square$

Lemma 14.2. *If a UPC graph contains chords of deficiency 1 and 2, then C does not contain a chord of deficiency 3.*

Proof. Suppose C_1, C_2, and C_3 are chords of deficiency 1, 2, and 3, respectively. Then by Lemma 14.1, C_1 is not skew to C_2, so the cycle formed by omitting the deficits of C_1 and C_2 has length $n - 3$. But the cycle formed by omitting the deficit of C_3 also has length $n - 3$, a contradiction. $\qquad\square$

Lemma 14.3. *Let G consist of a Hamilton cycle and two chords which are skew. Then G is not a UPC graph.*

Proof. By assumption, G has precisely 7 cycles. If G is a UPC graph, then it must have 9 vertices. By Lemma 14.1, G has no chord of deficiency 2. Hence, G must have chords with deficits 1 and 3. Now there is no 4-cycle, a contradiction. $\qquad\square$

Lemma 14.4. *Assume G has three chords and at least two chords are skew. If G is a UPC graph, then it must contain chords of deficit 1 and 2.*

Proof. Assume G has no chord of deficit 2. Then G must have a 4-cycle, say K, containing at least two chords. We consider the following two cases:

Case 1. The cycle K has precisely two chords, say C_1 and C_2. Since G has only one Hamilton cycle, it follows that C_1 and C_2 are not skew. We claim that C_3 is skew to both C_1 and C_2. Assume C_3 is skew to C_1 only (the case C_3 is skew to C_2 only is similar). Let H be the graph obtained from G by removing the deficits of C_2. Then C_1 has deficit 2 in H. An argument similar to that described in the proof of Lemma 14.1 shows that H must have two cycles of the same length, a contradiction. Hence, C_3 is skew to both C_1 and C_2. It is now easy to see that G has two cycles of the same length, a contradiction.

Case 2. The cycle K contains precisely three chords, say C_1, C_2, and C_3. Since G is not outerplanar, without loss of generality, we may assume C_1 is skew to C_3. Now the graph obtained from G by removing the deficit of C_2 has the chords C_1 and C_3 which are skew. It is easy to see that this graph has two cycles of the same length, a contradiction.

We now prove G that has a chord of deficit 1. Let C_1, C_2, and C_3 be three chords in G. Then G has a chord, say $C_1 = (v_2, v_3)$, of deficit 2. The chord C_1 cannot be skew to any other chord by Lemma 14.1. Hence, C_2 is skew to C_3. If G has no chord of deficit 1, then the 3-cycle in G must have precisely two chords, say C_1 and C_2. In addition, the $(n-1)$-cycle contains both $C_2 = (v_1, v_3)$ and $C_3 = (v_2, v_4)$. Then there must be precisely one vertex between vertices v_3 and v_4 in order to have a $v-1$-cycle. This forces G to have two 4-cycles, a contradiction.   □

Theorem 14. *A graph G with v vertices and v + 3 edges is UPC if and only if G is one of the graphs shown in Fig. 5.3.*   □

Proof. It is easy to see that the graphs given in Fig. 5.3 are all UPC. Now assume G is a graph with 3 chords, say C_1, C_2, and C_3. By Lemma 14.4, G contains one chord of deficit 1, say C_1, and one chord of deficit 2, say C_2. By Lemma 14.1, C_2 is not skew to any chord, so C_1 must be skew to C_3. By Lemma 14.2, G contains no chord of deficit 3, therefore it must contain a 5-cycle containing two chords. These two chords cannot be skew, for otherwise G would contain two $(n-1)$ cycles,

Fig. 5.3 The three non-isomorphic uniquely pancyclic graphs of order 14

a contradiction. Hence, the 5-cycle contains C_2 and C_3. It now follows that G is isomorphic to one of the graphs displayed in Fig. 5.3.   □

Using a method similar to that described in the proof of Lemma 14.4 we prove that, see [30], if a UPC graph has 4 chords and at least two chords are skew, then the graph must have chords of deficits 1 and 2.

Lemma 15.1. *Every UPC graph with four chords and at least two crossing chords contains a chord of deficit 2.*

Proof. The proof is by contradiction. Let C_1, C_2, C_3, and C_4 be 4 chords in G. Assume G does not have a chord of deficit 2. Then the 4-cycle must have more than one chord. We consider the following three cases:

Case 1. The 4-cycle contains precisely 2 chords, say C_1 and C_2. If C_1 and C_2 are skew, then G has two Hamilton cycles, a contradiction. Therefore C_1 and C_2 are not skew. If C_3 or C_4 crosses C_1 or C_2, then G has a chord of deficit 2 by Lemma 14.4. Now assume C_3 and C_4 do not cross C_1 or C_2. By assumption, C_3 and C_4 must be skew and the $(n-2)$-cycle in G must have two crossing chords. Without loss of generality, we may assume C_1 does not enclose any chord. Let G^* be the graph obtained from G by removing the deficit of C_1 and let $n^* = |V(G^*)|$. Then C_2 is a chord of deficit 2 in G^*. This forces G^* to have two (n^*-2)-cycles, a contradiction.

Case 2. The 4-cycle contains precisely 3 chords C_1, C_2, and C_3. An argument similar to that described in Case 2 of Lemma 14.4 shows that no pair of chords in $\{C_1, C_2, C_3\}$ can be skew. Hence, at least one of the chords C_1, C_2, or C_3 does not cross C_4. We may assume C_1 does not cross C_4. Let G^* be the graph obtained from G by removing the deficit of C_1. By the proof of Lemma 14.4, G^* contains two cycles of the same length, a contradiction.

Case 3. The 4-cycle contains four chords C_1, C_2, C_3, and C_4. Then two chords in $\{C_1, C_2, C_3, C_4\}$ must be skew. It is now easy to see that G has two cycles of the same length, a contradiction.

This completes the proof.   □

Lemma 15.2. *Every UPC graph with 4 chords and at least two crossing chords contains a chord of deficit 1.*

Proof. The proof is by contradiction. Let C_1, C_2, C_3, and C_4 be 4 chords in G. Assume G does not have a chord of deficit 1. Then the 3-cycle must have more than one chord. We consider the following two cases:

Case 1. The 3-cycle contains precisely two chords, say C_1 and C_2. If one of the chords C_1 or C_2 is of deficit 2, say C_1, then the two chords C_3 and C_4 cannot be skew to C_1 by Lemma 15.1. Since G has no chord of deficit 1, it follows that the $(n-1)$-cycle must have two crossing chords. Let $C_1 = v_2v_3$ and let G^* be the graph obtained from G by removing the deficit of C_1. Then C_2 in G^* is of deficit 1 and G^* has two (n^*-1)-cycles, one contains only chord C_2 and the other one is obtained from the $(n-1)$-cycle of G by replacing the arc enclosed by C_1 with C_1, a contradiction. Hence, none of C_1 and C_2 can be of deficit 2. We may now assume C_4 is of deficit 2. By Lemma 15.1, none of C_1, C_2, or C_3 is skew to C_4. Since C_1 and C_2 cannot be skew, we have two subcases.

Subcase 1.1 One of the chords C_1 or C_2 is skew to C_3. Without loss of generality, we may assume C_1 is skew to C_3. Then C_2 does not enclose any of the chords C_1 or C_3. Let G^* be obtained from G by removing the deficit of C_2. Then C_1 is of deficit 1 in G^*, hence produces a cycle of length $n^* - 1$ in G^*, where $n^* = |V^*(G)|$. An $(n^* - 1)$-cycle with two crossing chords can also be formed from $(n - 1)$-cycle of G by replacing the arc enclosed by C_2 with C_2, a contradiction.

Subcase 1.2 Both C_1 and C_2 are skew to C_3. Then there are only two layouts for G:

(a) The chords are $C_1 = v_2v_7$, $C_2 = v_3v_7$, $C_3 = v_1v_4$, and $C_4 = v_5v_6$, or
(b) The chords are $C_1 = v_2v_7$, $C_2 = v_3v_7$, $C_3 = v_1v_6$, and $C_4 = v_4v_5$.

If G has layout (a), then the $(n - 1)$-cycle in G contains two skew chords and cannot have three chords. If the $(n - 1)$-cycle of G contains the chords C_1 and C_3, then there are no vertices between v_1 and v_7 or between v_3 and v_4 on the Hamilton cycle. Therefore C_2 and C_3 are in a Hamilton cycle, a contradiction. So the $(n - 1)$-cycle must contain C_2 and C_3. This forces G to have exactly one vertex, say u, between v_1 and v_7 or between v_3 and v_4 on the Hamilton cycle. Now G contains two cycles of the same length, a contradiction.

If G has layout (b), then the $(n - 1)$-cycle of G cannot contain four chords. If the $(n - 1)$-cycle of G contains three chords, then there must be precisely one vertex between v_1 and v_7 or between v_6 and v_7 on the Hamilton cycle. This forces G to have two 4-cycles, a contradiction. If the $(n - 1)$-cycle of G contains only two chords, then there is no vertex between v_1 and v_2 or between v_6 and v_7 on the Hamilton cycle. This forces G to have two cycles of the same length, a contradiction.

Case 2. The 3-cycle has precisely three chords, say C_1, C_2, and C_3. Then one of these chords, say C_1, has deficit 2 by Lemma 14.4. If only one of the chords C_2 or C_3 is skew to C_4, then it is easy to see that G contains no $(n-1)$-cycle. Therefore C_2 and C_3 must be skew to C_4. Then G has the following layout: $C_1 = v_1v_5$, $C_2 = v_3v_5$, $C_3 = v_1v_3$, and $C_4 = v_2v_4$. Obviously, the $(n-1)$-cycle cannot contain three chords with two skew chords. If the $(n - 1)$-cycle contains C_2 and C_4, then there must be precisely one vertex between v_2 and v_3 or between v_4 and v_5 on the Hamilton cycle. This forces G to have two 5-cycles. Similarly, if the $(n - 1)$-cycle of G contains the chords C_3 and C_4, then G has two 5-cycles. This completes the proof. □

Theorem 15 ([30]). *There is no UPC graph with n vertices and $n + 4$ edges.*

Proof. Let G be a graph with n vertices and $n + 4$ edges. If G is outerplanar, then G cannot be a UPC graph by Theorem 14.2. Now assume G is not an outerplanar graph. We prove that G is not a UPC graph. The proof is by contradiction. Let G be a UPC graph. By Lemmas 15.1 and 15.2, G contains one chord of deficit 1, say C_1, and one chord of deficit 2, say C_2. By Lemma 15.1, C_2 cannot be skew to any other chord. Let C_3 and C_4 be the other two chords and let λ be the number of chords which are skew to C_1. Obviously, $\lambda \leq 2$.

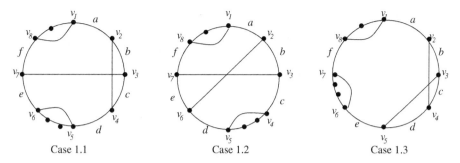

Fig. 5.4 C_1 does not skew to any chord; C_3 skews C_4

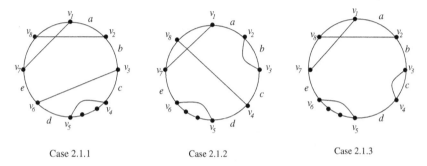

Fig. 5.5 C_1 skews C_3 only; C_3 does not skew C_4

If $\lambda = 0$, C_3 and C_4 must be skew. Therefore there are precisely 3 layouts for G, see Fig. 5.4.

If $\lambda = 1$, then C_1 skews to precisely one chord, say C_3. If C_4 does not skew to C_3, we obtain 3 layouts for G, see Fig. 5.5.

If C_3 and C_4 are skew and C_1 and C_4 are not adjacent, we obtain 2 layouts for G, see Fig. 5.6.

If C_3 and C_4 are skew and C_1 and C_4 are adjacent, there are only 3 layouts for G, see Fig. 5.7.

Finally, if $\lambda = 2$, then C_1 skews to both C_3 and C_4 and there are only 2 layouts for G, see Fig. 5.8.

Table 5.1 displays the number of cycles in each layout.

Let $M = \{\ell \mid \ell$ is the length of a cycle in $G\}$, $M^* = \{3, 4, 5, \ldots, n\}$, where n is the order of G, and

$$s =. \sum_{\ell \in M} \ell \tag{5.1}$$

$$s^* = (n + 3)(n - 2)/2. \tag{5.2}$$

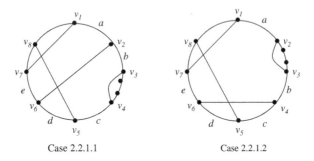

Case 2.2.1.1   Case 2.2.1.2

Fig. 5.6 C_3 skews C_4; C_1 and C_4 are not adjacent

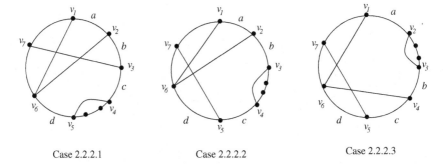

Case 2.2.2.1   Case 2.2.2.2   Case 2.2.2.3

Fig. 5.7 C_3 skews C_4; C_1 and C_4 are adjacent

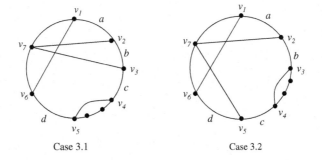

Case 3.1   Case 3.2

Fig. 5.8 C_1 skews to both C_3 and C_4

We investigate each case separately.

Case 1.1 A layout for G is given in Fig. 5.4. The chord $v_1 v_8$ is of deficit 1 and the chord $v_5 v_6$ is of deficit 2. The letters $a, b, c, d, e,$ *and* f indicate the number of edges between two consecutive vertices v_i's. The graph G has 19 cycles. Therefore its order must be 21 and

$$a + b + c + d + e + f = 16. \tag{5.3}$$

Table 5.1 Number of cycles in each layout

Case	1.1	1.2	1.3	2.1.1	2.1.2
Number of cycles	19	19	21	18	19
Case	2.1.3	2.2.1.1	2.2.1.2	2.2.2.1	2.2.2.2
Number of cycles	21	22	23	20	21
Case	2.2.2.3	3.1	3.2		
Number of cycles	21	21	21		

The lengths of the cycles in G are

$$M = \{3, 4, b + c + 1, c + d + e + 2, c + d + e + 4, b + d + e + 3,$$
$$b + d + e + 5, a + b + f + 2, a + b + f + 3, a + d + e + f + 3,$$
$$a + d + e + f + 4, a + d + e + f + 5, a + d + e + f + 6,$$
$$a + c + f + 3, a + c + f + 4, 18, 19, 20, 21\}.$$

By (5.1), (5.2), and (5.3) one can obtain $s = 3(a + d + e + f) + 210$ and $s^* = 228$. Since G is UPC, it follows that $s = s^*$, therefore $a + d + e + f = 6$. Now by (5.3) we obtain $b + c = 10$. Therefore

$$M = \{3, 4, 11, c + d + e + 2, c + d + e + 4, b + d + e + 3,$$
$$b + d + e + 5, a + b + f + 2, a + b + f + 3, 9, 10, 11, 12,$$
$$a + c + f + 3, a + c + f + 4, 18, 19, 20, 21\}.$$

This is a contradiction because G has two 11-cycles.

Case 1.2 A layout for G is given in Fig. 5.4. The order of G is 21 and $a + b + c + d + e + f = 16$. It is straightforward to see that $s = 3(a + c + d + f) + 211$ and $s^* = 228$. Since G is UPC, it follows that $s = s^*$ and $3(a + c + d + f) = 17$, a contradiction.

Case 1.3 The order of G is 23 and $a + b + c + d + e + f = 18$, see Fig. 5.4. In this case we obtain $s = 6(a + e + f) + 263$, $s^* = 273$, and $s = s^*$. This leads to $6(a + e + f) = 10$, a contradiction.

Case 2.1.1 The order of G is 20 and $a + b + c + d + e = 15$, see Fig. 5.5. For this case $s = 4(b + e) + 2(c + d) + 182$, $s^* = 207$, and $s = s^*$. This leads to $2(2b + 2e + c + d) = 25$, a contradiction.

Case 2.1.2 The order of G is 21 and $a + b + c + d + e = 16$, see Fig. 5.5. We see that $s = 5(a + c + d + e) + 182$, $s^* = 228$, and $s = s^*$. This leads to $5(a + c + d + e) = 46$, a contradiction.

Case 2.1.3 See Fig. 5.5. We see that

$$M = \{3, 4, c + 1, a + b + d + e + 3, a + b + d + e + 4,$$
$$a + b + d + e + 5, a + b + d + e + 6, b + d + e + 4, b + d + e + 5,$$
$$b + d + e + 6, b + d + e + 7, b + c + d + e + 3, b + c + d + e + 4,$$
$$b + c + d + e + 5, b + c + d + e + 6, a + 2, a + 3, 20, 21, 22, 23\},$$

where

$$a + b + c + d + e = 18. \tag{5.4}$$

Now by (5.1) and (5.2) we obtain $s = 7(b + d + e) + a + 247 = 273 = s^*$, hence, $7(b + d + e) + a = 26$. This leads to $b + d + e \leq 3$ and $a \geq 5$, therefore $b + d + e + 4 \leq a + 2$. In addition, by (5.4), $6a + 7c = 100$, hence $c + 1 > 5$. It is now easy to see that $b + d + e + 4 = \min M \setminus \{3, 4\}$. This forces $b + d + e + 4 = 5$, that is, $b + d + e = 1$, hence, $a = 19$. This contradicts (5.4).

Case 2.2.1.1 See Fig. 5.6. The order of G is 24 and $a + b + c + d + e = 19$. It is easy to see that $s = 2(a + 2b + 2c + e) + 274$, $s^* = 297$, and $s = s^*$. This leads to $2(a + 2b + 2c + e) = 23$, a contradiction.

Case 2.2.1.2 See Fig. 5.6. We have

$$M = \{3, 4, c + d + 1, d + e + 2, d + e + 3, c + e + 3, c + e + 4,$$
$$a + b + e + 3, a + b + e + 4, a + b + e + 5, a + b + e + 6,$$
$$a + b + c + 3, a + b + c + 4, a + b + c + 5, a + b + c + 6,$$
$$a + b + d + 4, a + b + d + 5, a + b + d + 6, a + b + d + 7,$$
$$22, 23, 24, 25\},$$

where

$$a + b + c + d + e = 20. \tag{5.5}$$

A simple calculation shows that $s = 5(a + b) + e + 312$ and $s^* = 322$. Hence, $5(a + b) + e = 10$. Since $v_6 \neq v_7$, it follows that $e \geq 1$, therefore $a + b \leq 1$. If $a + b = 1$, then $e = 5$. Hence, $d + e + 2 = a + b + d + 6 = d + 7$. Therefore G contains two $(d+7)$-cycles, a contradiction. If $a + b = 0$, then $e = 10$. Rewriting M with these values we have

$$M = \{3, 4, 11, d + 12, d + 13, c + 13, c + 14, 13, 14, 15, 16, c + 3,$$
$$c + 4, c + 5, c + 6, d + 4, d + 5, d + 6, d + 7, 22, 23, 24, 25\}.$$

Therefore, we must have either $c + 3 = 5$, which implies $c + 13 = 15$ or $d + 4 = 5$, which implies $d + 12 = 13$. Hence, G has two cycles of the same length, a contradiction.

Case 2.2.2.1 See Fig. 5.7. We see that

$$M = \{3, 4, a + 2, a + 3, b + 3, b + 4, c + d + 3, c + d + 4, c + d + 5,$$
$$c + d + 6, b + c + d + 2, b + c + d + 4, a + b + 2, a + b + 3,$$
$$a + c + d + 4, a + c + d + 6, 19, 20, 21, 22\},$$

where $a + b + c + d = 17$. Therefore by (5.1) and (5.2), $s = 2c + 2d + 242 = 250 = s^*$. Hence, $c + d = 4$ and $a + b = 13$. Rewriting M, we obtain

$$M = \{3, 4, a + 2, a + 3, b + 3, b + 4, 7, 8, 9, 10, b + 6,$$
$$b + 8, 15, 16, a + 8, a + 10, 19, 20, 21, 22\}.$$

If $a \leq 3$, then G has two cycles of the same length. Therefore, $a \geq 4$. Now G is missing a 5-cycle, a contradiction.

Case 2.2.2.2 See Fig. 5.7. The set of cycle lengths is

$$M = \{3, 4, a + 2, a + 3, d + 2, d + 3, a + d + 3, b + c + 4, b + c + 5,$$
$$b + c + 6, b + c + 7, a + b + c + 3, a + b + c + 4,$$
$$a + b + c + 5, a + b + c + 6, b + c + d + 2, b + c + d + 4,$$
$$20, 21, 22, 23\}.$$

Therefore $s = 152 + 7a + 10b + 10c + 5d$. We also have $a + b + c + d = 18$, $s^* = 273$, and $s = s^*$. Hence, $2a + 5(b + c) = 31$. We now obtain $(a, b + c, d) \in \{3, 5, 10), (8, 3, 7), (13, 1, 4)\}$. In each case G has two cycles of the same length, a contradiction.

Case 2.2.2.3 See Fig. 5.7. The set of cycle lengths is

$$M = \{3, 4, d + 2, d + 3, c + 3, c + 4, a + b + 3, a + b + 4,$$
$$a + b + 5, a + b + 6, c + d + 1, a + b + c + 3, a + b + c + 4,$$
$$a + b + c + 5, a + b + c + 6, a + b + d + 4, a + b + d + 6,$$
$$20, 21, 22, 23\}.$$

Therefore $s = 152 + 10a + 10b + 7c + 5d$. We also have $a + b + c + d = 18$, $s^* = 273$, and $s = s^*$. Hence, $2c + 5(a + b) = 31$. It follows that $(a + b, c, d) \in \{(5, 3, 10), (3, 8, 7), 1, 13, 4)\}$. In each case G has two cycles of the same length, a contradiction.

Case 3.1 See Fig. 5.8. The set of cycle lengths is

$$M = \{3, 4, a + 2, a + 3, b + 2, a + b + 2, a + b + 3, a + c + d + 4,$$
$$a + c + d + 6, b + c + d + 3, b + c + d + 4, b + c + d + 5,$$
$$b + c + d + 6, c + d + 3, c + d + 4, c + d + 5,$$
$$c + d + 6, 20, 21, 22, 23\}.$$

Therefore $s = 6a + 7b + 10c + 10d + 151$. On the other, $a + b + c + d = 18$, $s^* = 273$, and $s = s^*$. Hence, $b + 4(c + d) = 14$. So $(a, b, c + d) \in \{(13, 2, 3), (10, 6, 2), (7, 10, 1)\}$. In every case G has two cycles of the same length.
Case 3.2 See Fig. 5.8. The set of cycle lengths is

$$M = \{3, 4, a + 2, a + 3, d + 2, d + 3, a + d + 3, b + c + 3, b + c + 5,$$
$$a + b + c + 3, a + b + c + 4, a + b + c + 5, a + b + c + 6,$$
$$b + c + d + 3, b + c + d + 4, b + c + d + 5, b + c + d + 6,$$
$$20, 21, 22, 23\}.$$

Therefore $s = 150 + 7a + 10b + 10c + 7d$. On the other, $a + b + c + d = 18$, $s^* = 273$, and $s = s^*$. Hence, $276 + 3(b + c) = 273$, a contradiction.   □

Markström [23] verified that the graphs given in Figs. 5.2 and 5.3 are the only UPC graphs with at most 5 chords. This supports the following conjecture of Shi [30]:

Conjecture 5.1. There is no UPC graph with n vertices and $n + m$ edges for $m \geq 4$.

5.5 Cycle Space of a Graph

The material in this section can also be found in [33]. Some proofs are found in [37], although using the language of matroids. A graph-theoretic version may be found in [7], although without proof; both [7, 33] refer to [15], which proves them.

Consider a graph G. We shall write E for the edge set $E(G)$ of G. Write P_E for the set of all subsets of E, including the empty set. Define the binary operation $\oplus$ on P_E as follows: $A \oplus B = (A \setminus B) \cup (B \setminus A)$, the *symmetric difference of A and B*, where $A, B \in P_E$. Obviously, P_E is closed under $\oplus$. Furthermore,

1. $(A \oplus B) \oplus C = A \oplus (B \oplus C)$ for every $A, B, C \in P_E$;
2. $A \oplus \emptyset = A$ for every $A \in P_E$;
3. $A \oplus A = \emptyset$ for every $A \in P_E$;
4. $A \oplus B = B \oplus A$ for every $A, B \in P_E$;

Hence, $(P_E, \oplus)$ is an abelian group.

Consider GF(2), the finite field of order 2. Define the operation $* : GF(2) \times P_E \to P_E$ by $1 * A = A$ and $0 * A = \emptyset$, where $A \in P_E$ and $0, 1 \in GF(2)$. Then for $\alpha, \beta \in GF(2)$ and $A, B \in P_E$ we have

1. $(\alpha + \beta) * A = (\alpha * A) \oplus (\beta * B)$;
2. $\alpha * (A \oplus B) = (\alpha * A) \oplus (\alpha * B)$;
3. $(\alpha \cdot \beta) * A = \alpha * (\beta * A)$.

Hence, $(P_E, \oplus)$ is a vector space over $GF(2)$. We note that

1. Every member of P_E can be written as a linear combination of the edges over GF(2) and
2. The set of edges of G are independent over GF(2).

Theorem 16. *Let G be a graph with e edges. Then $(P_E, \oplus)$ is an e-dimensional vector space over GF(2).*

We now study a particular subspace of the vector space $(P_E, \oplus)$ over GF(2), called the *cycle space* of G. Let P_C be the set of all cycles and the unions of the edge-disjoint cycles of G, including the empty set.

We make use of the following theorem in this section:

Theorem 17. *A graph H can be expressed as the union of the edge-disjoint cycles if and only if every vertex in H is of even degree.*

Proof. Clearly, if the graph H can be expressed as the union of the edge-disjoint cycles, then every vertex in H is of even degree. Conversely, assume every vertex of H is of even degree. We prove that H can be expressed as the union of the edge-disjoint cycles. If every vertex of H has degree zero, then the statement is obviously true. Otherwise, since every vertex of H is of even degree, it follows that H is not a forest. Therefore H contains a cycle, say C_1. Let $H_1 = H \setminus C_1$. Then every vertex of H_1 is of even degree. If H_1 has no edges, then $H = C_1$ and the statement is true. Otherwise, H_1 contains a cycle, say C_2. Let $H_2 = H_1 \setminus C_2 = (H \setminus C_1) \setminus C_2$. If H_2 has no edges, then $H = C_1 \cup C_2$ and we are done. Otherwise we repeat this process. Since H is a finite graph, we will obtain $H = C_1 \cup C_2 \cup \ldots \cup C_r$, where C_i is a cycle for $1 \leq i \leq r$ and $r \geq 1$. This completes the proof. $\square$

Theorem 18. *The cycle space $(P_C, \oplus)$ is a subspace of the vector space $(P_E, \oplus)$ over GF(2).*

Proof. It suffices to show that P_C is closed under operation $\oplus$. Note that every member of P_C is a subgraph of G with all vertices of even degree. Let $C_1, C_2 \in P_C$ and $C_3 = C_1 \oplus C_2$. We show that $C_3 \in P_C$. If $C_1 = C_2$, then $C_3 = \emptyset \in P_C$. If $C_1 = \emptyset$ or $C_2 = \emptyset$, then $C_3 = C_2$ or $C_3 = C_1$, respectively, hence, $C_3 \in P_C$. Now assume $C_1 \neq \emptyset$, $C_2 \neq \emptyset$ and $C_1 \neq C_2$. Pick a vertex x in C_3. Obviously, x is a vertex of C_1 or C_2. Let A_i be the set of edges at vertex x in C_i. Since $C_3 = C_1 \oplus C_2$, it follows that $A_3 = A_1 \oplus A_2$. Hence, $|A_3| = |A_1| + |A_2| - 2|A_1 \cap A_2|$. By assumption, $|A_1|$ and $|A_2|$ are even, so $|A_3|$ is even. Therefore every vertex in C_3 is of even degree. Hence, $C_3 \in P_C$ by Theorem 17. $\square$

In what follows we find the dimension of the cycle space, P_C, of a connected graph G with n vertices and e edges. Let T be a spanning tree of G. The edges of T are called *branches* and the edges of $E(G) \setminus E(T)$ are called *links*. We label the links by $\ell_1, \ell_2, \ldots, \ell_{e-n+1}$. The *fundamental* cycle with respect to the link ℓ_i is the unique cycle C_i in the graph $T \cup \ell_i$. The set $\{C_i \mid i = 1, 2, \ldots, e - n + 1\}$ is the set of *fundamental cycles* of G with respect to T. We note that

1. Every fundamental cycle C_i contains precisely one link, namely ℓ_i.
2. The link ℓ_i is present only in the fundamental cycle C_i with respect to T.

Therefore the edge set of no fundamental cycle can be written as a linear combination of some or all of the remaining fundamental cycles. In other words, the fundamental cycles C_i, $i = 1, 2, \ldots, e - n + 1$ are independent. We now prove that every subgraph in the cycle space P_C can be written as a linear combination of C_i's.

Pick a subgraph C in P_C. Let $\ell_{i_1}, \ell_{i_2}, \ldots, \ell_{i_r}$ be the only links in C. Set $C' = C_{i_1} \oplus C_{i_2} \oplus \ldots \oplus C_{i_r}$. Clearly, C' contains the links $\ell_{i_1}, \ell_{i_2}, \ldots, \ell_{i_r}$ and no other links are in C'. Since P_C is a vector space, $C \oplus C' \in P_C$. Therefore if $C \oplus C' \neq \emptyset$, it must contain a cycle. On the other hand, $C \oplus C'$ consists of branches and hence it has no cycle, a contradiction. Thus we must have $C \oplus C' = \emptyset$, which implies $C = C'$.

We are now ready to state the main theorem of this section.

Theorem 19. *Let G be a connected graph with n vertices and e edges. Then the fundamental cycles with respect to a spanning tree of G is a basis for the cycle space P_C of G. Hence, the dimension of P_C is $e - n + 1$.*

Corollary 19.1. *Let G be a UPC graph with k chords. Then the dimension of the cycle space of G is $k + 1$. Hence, the cycle space of G has 2^{k+1} members, including the empty set.*

If G is not connected, the union of the fundamental cycles for the connected components of G forms a basis for the cycle space of G.

Corollary 19.2. *Let G be a graph with n vertices and e edges, having p connected components. Then the dimension of the cycle space of G is $e - n + p$.*

5.6   Bounds on the Number of Edges in a UPC Graph

The material in this section is based on a paper of Markström published in 2009 (see [23]). A UPC graph on n vertices consists of a Hamilton cycle C_n and some chords. In this section we use k for the number of chords in a UPC graph, c for the number of skew pairs of chords, c_3 for the number of unordered triples of pairwise skew chords, and c_Δ for the number of unordered triples $\{e_1, e_2, e_3\}$ of chords, no two skew, such that there is a cycle containing all three chords.

Let k be the number of chords in a UPC graph G and let Δ be its maximum degree. We define the function f as follows:

$$f(k, \Delta) = \begin{cases} \frac{1}{24}(8k^2 - 32k + 23) & \text{for } \Delta = 3 \\ 7.2^{k-2}\left(1 - \frac{1}{14}\left(1 + \Delta + \frac{1}{2}\binom{\Delta}{2}\right)\right) & \text{for } \Delta \geq 4. \end{cases}$$

The following theorem gives upper bounds for the number of chords in a UPC graph:

Theorem 20 ([23]). *Let G be a UPC graph of order n with k chords. Then*

$$3 + 2k + \binom{k}{2} + c(k - 1) - c_3 + c_\Delta \leq n. \tag{5.6}$$

In addition, if $k \geq 4$, then

$$\log_2\left(n - 1 + f(k, \Delta)\right) + \log_2\left(\frac{4}{7}\right) \leq k. \tag{5.7}$$

Proof. The number of cycles in G with no chords is one and with one chord is $2k$. A pair of skew chords are in two cycles and a pair of nonskew chords are in one cycle. This gives us $1 + 2k + \binom{k}{2} + c$ cycles. Now we count the cycles containing precisely three chords.

1. The number of cycles with a set of three chords, no two skew, is c_Δ.
2. For each set of three chords containing two skew chords and a third chord not skew to either of them there is exactly one cycle containing precisely the three chords.
3. For each set of three chords containing two nonskew chords and one chord skew to both of them, there are two cycles containing the three chords.
4. For each set of three pairwise skew chords, there are two cycles containing the three chords.

When three chords are used the number of cycles containing a given triple in cases 2 and 3 is the same as the number of skew pairs in the triple. In case 4, we obtain two cycles when there are three skew pairs. Since there are k chords, c skew pairs of chords and c_3 unordered triples of pairwise skew chords, it follows that there are at least $c(k - 2) - c_3$ cycles containing exactly three chords with at least one pair skew. There may also be cycles containing four or more chords. Hence, the total number of cycles in G is at least $1 + 2k + \binom{k}{2} + c(k - 1) - c_3 + c_\Delta$. On the other hand, G has precisely $n - 2$ cycles. This proves inequality (5.6).

In order to prove the second inequality we make use of the cycle space of graph G. Since G is a UPC graph with precisely k chords, its cycle space has dimension $k + 1$ by Corollary 19.1. Therefore this cycle space has 2^{k+1} members, including the empty graph. Let a be the number of members which are not cycles and are different from the empty graph. These members are called a-graphs. Any a-graph is a subgraph of G whose vertices are even degree and contains a vertex of degree greater than 3 and/or several components.

Since G has precisely $n - 2$ cycles, it follows that

$$2^{k+1} - 1 - a = n - 2. \tag{5.8}$$

We now prove that $a \geq 2^{k-2} + f(k, \Delta)$. Each chord makes two cycles. For a chord e we denote the shorter cycle by C_e. If the two cycles have the same length, C_e is taken as the cycle containing the edge $\{1, n\}$. We now consider two cases.

Case 1: $\Delta(G) \geq 4$. Let v be a vertex of maximum degree in G. A subset of size at least three of the chords at v, together with any subset of the chords not at c produce an a-graph. Adding the Hamilton cycle to this a-graph gives another a-graph. Similarly, a subset of size 2 of the chords at v, together with any subset of the chords not at v produce an a-graph directly or after adding the Hamilton cycle to this subgraph. Therefore the number of nontrivial a-graphs is at least

$$2\left(2^\Delta - \left(1 + \Delta + \binom{\Delta}{2}\right)\right)2^{k-\Delta} + \binom{\Delta}{2}2^{k-\Delta}$$
$$= 2^{k+1} - 2^{k+1-\Delta}\left(1 + \Delta + \tfrac{1}{2}\binom{\Delta}{2}\right)$$
$$= 2^{k-2} + 7.2^{k-2}\left(1 - \tfrac{2^{3-\Delta}}{7}\left(1 + \Delta + \tfrac{1}{2}\binom{\Delta}{2}\right)\right)$$
$$\geq 2^{k-2} + 7.2^{k-2}\left(1 - \tfrac{1}{14}\left(1 + \Delta + \tfrac{1}{2}\binom{\Delta}{2}\right)\right).$$

Hence, $a \geq 2^{k-2} + f(k, \Delta)$.

Case 2: $\Delta(G) = 3$. Then the unique 3-cycle in G contains only one chord, say e_1. This 3-cycle is denoted C_{e_1}. There may be a chord, say e_2, at the third vertex of C_{e_1}. Let A be a set consists of the other chords. Obviously, $A \neq \emptyset$. Define $C_A = \sum_{e \in A} C_e$. Now either C_A or $C_A + C_{e_1} + C_v$ is an a-graph. If there is no chord e_2, we will obtain $2^{k-1} - 1 = 2^{k-2} + 2^{k-2} - 1$ a-graphs. Hence, $a \geq 2^{k-1} - 1 \geq 2^{k-2} + f(k, \Delta)$, where $\Delta = 3$. If there is a chord e_2, we will obtain 2^{k-2} a-graphs this way.

Now assume there is a chord e_2. We will obtain more a-graphs as follows. Let m be the number of chords which do not cross e_2. Any such chord together with e_2 (or e_1 and e_2) produce one a-graph with two components. Similarly, any pair of chords which are not crossing e_2 together with e_2 (or e_1 and e_2) give one a-graph with two components. This way we get $2(m + \binom{m}{2})$ a-graphs.

Let ℓ be the number of chords, apart from e_1, which cross e_2. Any pair of such chords, if they cross each other, together with e_1 and e_2 form one a-graph with two components, and if they do not cross each other, also form one a-graph with two components which contain neither e_1 nor e_2. This gives us $\binom{\ell}{2}$ a-graphs. Therefore when there is a chord e_2 we obtain at least

$$2^{k-2} + 2\left(m + \binom{m}{2}\right) + \binom{\ell}{2}$$

a-graphs.

Since $m + \ell = k - 2$, it follows that

$$2\left(m + \binom{m}{2}\right) + \binom{\ell}{2} = \frac{1}{2}\left(3m^2 + (7 - 2k)m + k^2 - 5k + 6\right)$$

$$\geq \frac{1}{24}(8k^2 - 32k + 23).$$

Hence, $a \geq 2^{k-2} + f(k, \Delta)$, where $\Delta = 3$.

Therefore in all cases we have $a \geq 2^{k-2} + f(k, \Delta)$. Now by (5.8) we obtain $2^{k+1} - 2^{k-2} - f(k, \Delta) \geq n - 2$. This implies inequality (5.7).   □

5.7   Open Problems

In 1975, Erdös asked for the maximum number of edges in a graph with no repeated cycle lengths. This question is one of the open problems mentioned in [3]. Lai [22] using only cycles of length $0(n)$ proves that the maximum number of edges in a graph with no repeated cycle lengths is asymptotically at least $n + \sqrt{2.4n}$, which is greater than the bound given in inequality (5.7). Note that having a Hamilton cycle in graph G is essential in the proof of this inequality. Finally, the following problem is found in [23]: Determine the maximum number of edges in a Hamiltonian graph on n vertices with no repeated cycle lengths.

Chapter 6
Bipancyclic Graphs

6.1 Introduction

Recall that a *bipartite graph* is a graph with two disjoint sets of vertices, V_1 and V_2 say, where no two vertices in the same set are adjacent, and is a *balanced bipartite graph* if V_1 and V_2 are equal in size. Obviously a bipartite graph cannot be pancyclic, as it can contain no odd cycles. So, in 1982, Schmeichel and Mitchem [29] defined a *bipancyclic* graph G to be a balanced bipartite graph that contains cycles of all even orders from 4 up to and including the number of vertices of G. A *minimally* bipancyclic graph is one with the smallest possible number of edges, given its number of vertices, and a *uniquely* bipancyclic graph is one with exactly one cycle of every possible order. The complete bipartite graph $K_{n,n}$ is obviously bipancyclic when $n \geq 2$.

Suppose G is a bipancyclic graph on $2n$ vertices. Then G contains a Hamilton cycle, say $C = (a_1, a_2, \ldots, a_{2n}, a_1)$, and every second vertex in C must be in the same set, V_1 or V_2. If two vertices x and y are adjacent, then one of i and j must be even and the other odd. It follows that V_1 and V_2 must be the same size, that is G is a balanced bipartite graph. We write $n = |V_1| = |V_2|$. We shall follow this notation throughout this chapter.

6.2 Edge Number Conditions

Theorem 21 ([10]). *Suppose G is a balanced bipartite graph on $2n$ vertices. If G has more than $n(n-1) + 1$ edges, then G is bipancyclic.*

Proof. Case $n = 1$ is impossible. If $n = 2$, G has 4 edges and is $K_{2,2}$, which is trivially bipancyclic. We proceed by induction on n. If G is $K_{n,n}$ we are finished, so assume there is a vertex x in V_1 whose degree $d(x)$ is less than n. If $d(x) = 0$, then

© The Author(s) 2016
J.C. George et al., *Pancyclic and Bipancyclic Graphs*, SpringerBriefs
in Mathematics, DOI 10.1007/978-3-319-31951-3_6

G has at most $n^2 - n$ edges, so $d(x) \geq 1$. Select $y \in V_2$ adjacent to x. Now consider the graph G_1 formed by deleting x and y from G. We have deleted at most $(2n - 2)$ edges (at most $n - 1$ containing x and n containing y, and (x, y) is counted twice) so the number of edges in G_i is more than $n(n-1) + 1 - (2n-2) = (n-1)(n-2) + 1$, so by induction G_1 is bipancyclic. So G contains cycles of all even lengths up to $2n - 2$.

It remains to show that G contains a Hamilton cycle. We know that G_i contains a Hamilton cycle, C say, with every second vertex in V_1 and every other vertex in V_2. If possible, select an edge (z, t) in C, where $z \in V_1, t \in V_2$. If (x, t) and (y, z) are edges in G, then we obtain a Hamilton cycle in G by deleting edge (z, t) and adding the path (z, y, x, t). If no such edge has this property, then for each edge (z, t) in C ($z \in V_1, t \in V_2$) one of $\{(x, t), (y, z)\}$ is not an edge of G. So $n - 1$ of the possible edges in G are not present, and $|E(G)| \leq n^2 - (n - 1) \leq n(n - 1) + 1$, a contradiction. $\square$

The above theorem is best possible. To see this, consider a balanced bipartite graph formed from $K_{n,n-1}$ by adding a vertex adjacent to exactly one of the vertices in the larger component set. This graph is not bipancyclic, because it is not Hamiltonian, but it is a balanced bipartite graph on $2n$ vertices, with $n(n - 1) + 1$ edges.

If we assume the balanced bipartite graph is Hamiltonian, a better result is available.

Theorem 22 ([10]). *Suppose G is a Hamiltonian bipartite graph on $2n$ vertices, $n > 3$. If G has more than $n^2/2$ edges, then G is bipancyclic.*

(It is easy to see that the requirement $n > 3$ is necessary; when $n = 3$, the requirement of more than $n^2/2$ edges is met by C_6, which contains no 4-cycle.)

Proof. The case $n = 4$ is easily verified. We proceed by induction on n. Suppose G has components V_1 and V_2 and contains a Hamilton cycle $C = (a_1, a_2, \dots, a_{2n}, a_1)$, where the odd-numbered vertices are in V_1. Write E_k for the set of edges of the form $(a_{2i}, a_{2i+2k-1})$ (subscripts modulo $2n$). The sets $E_1, E_2, \dots, E_n$ contain every edge of G precisely once. Notice that E_1 consists of all the edges (a_{2i}, a_{2i+1})—every second edge of C—so $|E_1| = n$, and similarly $|E_n| = n$.

Suppose G has no $(2n - 2)$-cycle. If there is an edge (a_{2i}, a_{2i+3}), then $(a_1, a_2, \dots, a_{2i}, a_{2i+3}, \dots, a_{2n}, a_1)$ is a $(2n - 2)$-cycle, so E_2 must be empty, and so must be E_{n-1}. Moreover, G cannot contain both of the edges $(a_{2i}, a_{2i+2k-1})$ and $(a_{2i+2}, a_{2i+2k+1})$, so $|E_k| \leq n/2$ for $3 \leq k \leq n-2$. Therefore two of the sets E_i have n elements, two are empty, and the other $n - 4$ have at most $n/2$ elements. Adding, we obtain $E(G) \leq 2n + (n/2)(n - 4) = n^2/2$, contradicting the assumption.

Therefore G contains a $(2n - 2)$-cycle, say $D = (b_1, b_2, \dots, b_{2n-2}, b_1)$. Write b_{2n-1} and b_{2n} for the other two vertices of G, with b_{2n-1} in the same component as b_1. If the induced subgraph of G based on the vertices of D contains more than $(n - 1)^2/2$ edges, it is bipancyclic, and therefore G is bipancyclic and we are done. So let us assume that the induced subgraph has at most $(n - 1)^2/2$ edges, whence the number of edges of D containing b_{2n-1} or b_{2n} must be greater than $n^2/2 - (n-1)^2/2 = n - \frac{1}{2}$. So there are at least n edges incident with b_{2n-1} or b_{2n} (or both).

Suppose b_{2n} is adjacent to more than $(n-1)/2$ vertices of D. Select m such that $2 \leq m \leq n - 2$. Then for some odd i satisfying $i \leq 2n - 3$, b_{2n} must be adjacent to both b_i and b_{i+2m-2} (subscript reduced modulo $2n - 2$), and G contains the $2m$-cycle $(b_i, b_{i+1}, \ldots, b_{i+2m-2}, b_i)$. So G contains cycles of every even length from 4 to $2n - 4$. Together with C and D, this shows that G is bipancyclic. So, if G is not bipancyclic, b_{2n} is adjacent to at most $(n-1)/2$ vertices of D. The same argument applies to b_{2n-1}. Therefore, if G is not bipancyclic, the only way the required number of edges adjacent to b_{2n-1} or b_{2n} is if those two vertices are adjacent and each is adjacent to exactly $(n-1)/2$ vertices of D. To finish the proof, we assume that b_{2n-1} and b_{2n} satisfy this property.

Suppose G is missing a $2m$-cycle for some m satisfying $2 \leq m \leq n - 2$. From the description of b_{2n} it follows that for each odd i satisfying $1 \leq i \leq 2n - 3$ exactly one of (b_{2n}, b_i) and (b_{2n}, b_{i+2m-2}) is an edge of G. Similarly, exactly one of (b_{2n}, b_i) and (b_{2n}, b_{i+2m+2}) is an edge. So, if (b_{2n}, b_i) is not an edge, both (b_{2n}, b_{i+2m-2}) and (b_{2n}, b_{i+2m+2}) are edges (or perhaps they could be the same edge). But then (b_{2n-1}, b_{i-1}) is an edge, so G contains the $2m$-cycle $(b_{i-1}, b_{i-2}, \ldots, b_{i-2m+2}, b_{2n}, b_{2n-1}, b_{i-1})$. Similarly, (b_{2n-1}, b_{i+1}) is an edge, whence G contains the $2m$-cycle $(b_{i+1}, b_{i+2}, \ldots, b_{i+2m-2}, b_{2n}, b_{2n-1}, b_{i-1})$. Neither of these can happen if G is missing a $2m$-cycle.

But this implies that b_{2n-1} is adjacent to fewer vertices of D than b_{2n} is, contradicting the result found two paragraphs above. So the assumption that G is not pancyclic is false. □

6.3 Degree Conditions

From the preceding section, it is clearly useful to know whether the graph under consideration is Hamiltonian. So the following theorem, due to Chvátal, is important:

Theorem 23 ([8]). *Suppose G is a balanced bipartite graph on $2n$ vertices, $n > 3$, with components $V_1 = \{x_1, x_2, \ldots, x_n\}$ and $V_2 = \{y_1, y_2, \ldots, y_n\}$, where the degrees satisfy*

$$d(x_1) \leq d(x_2) \leq \ldots \leq d(x_n) \tag{6.1}$$

and

$$d(y_1) \leq d(y_2) \leq \ldots \leq d(y_n). \tag{6.2}$$

If

$$d(x_k) \leq k < n \Rightarrow d(y_{n-k}) \geq n - k + 1, \tag{6.3}$$

then G is Hamiltonian.

Proof. From the graph G, construct a new graph G^* by joining all vertices in V_2. Then G^* clearly satisfies the conditions of Theorem 4, so it is Hamiltonian. As no two vertices in V_1 are adjacent in G^*, and since they constitute half the vertices, every second vertex in the Hamilton cycle must be in V_1, so all the edges of the Hamilton cycle have one endpoint in V_1. So all the edges are edges of G. So G is Hamiltonian.   □

Obviously this theorem is also true if (6.3) is replaced by

$$d(y_k) \leq k < n \Rightarrow d(x_{n-k}) \geq n - k + 1. \tag{6.4}$$

Bipancyclic graphs were first defined by Schmeichel and Mitchem [29]. They then proved the following theorem:

Theorem 24 ([29]). *Suppose G is a bipartite graph on $2n$ vertices, $n > 3$, with disjoint vertex sets $V_1 = \{x_1, x_2, \ldots, x_n\}$ and $V_2 = \{y_1, y_2, \ldots, y_n\}$, where the degrees satisfy (6.1), (6.2), and (6.3). Then G is bipancyclic.*

The theorem is obviously true when $n = 2$, and false when $n = 3$ (consider C_6).

We shall follow the proof in [29], but break the theorem into two parts, Theorems 25 and 26.

Theorem 25. *Suppose G satisfies the conditions of Theorem 24. If G is not bipancyclic, then n is odd, G contains at least $n + 3$ vertices with degree $\frac{1}{2}(n + 1)$ or greater, and there is a Hamilton cycle $C = (a_1, a_2, \ldots, a_{2n}, a_1)$ in which two consecutive vertices have degree exactly $\frac{1}{2}(n + 1)$.*

The proof requires five lemmas.

Lemma 25.1 ([29]). *Suppose G is a bipartite graph containing a Hamilton cycle $C = (a_1, a_2, \ldots, a_{2n}, a_1)$.*

(i) *If $d(a_1) + d(a_{2n}) > n + 1$, then G is bipancyclic.*

(ii) *If $d(a_1) + d(a_{2n}) \geq n + 1$, and G is not bipancyclic, suppose G is missing a cycle of length $2m$. Then for each odd integer k, $3 \leq k \leq 2n - 1$, exactly one of the pairs (a_{2n}, a_k) and $(a_1, a_{f_{2m}(k)})$ is an edge of G, where $f_{2m}(k) = 2n - 2m + k + 1$ when $3 \leq k \leq 2m - 3$ and $f_{2m}(k) = k - 2m + 3$ when $2m - 1 \leq k \leq 2n - 1$.*

(Of course, the lemma applies if a_1 and a_{2n} are replaced by any two vertices that are consecutive in C.)

Proof. Let us call a vertex a_k as odd or even according as k is odd or even. Then a_1 is adjacent only to even vertices and a_{2n} only to odd ones.

We first observe that if both (a_{2n}, a_k) and $(a_1, a_{f_{2m}(k)})$ are edges of G, then G contains a cycle of length $2m$: when $3 \leq k \leq 2m - 3$, the cycle is $(a_1, a_2, \ldots, a_k, a_{2n}, a_{2n-1}, \ldots, a_{2n-2m+k+1}, a_1)$, while if $2m - 1 \leq k$, the cycle is $(a_1, a_{k-2m+3}, a_{k-2m+4}, \ldots, a_k, a_{2n}, a_1)$.

Now, as k ranges through the odd integers from 3 to $2m - 3$, $f_{2m}(k)$ covers the even integers from $2n - 2m + 4$ to $2n - 4$, and as k ranges through the odd integers from $2m - 1$ to $2n - 1$, $f_{2m}(k)$ covers the even integers from 2 to $2n - 2m + 2$, and there are no repeats.

As $a_{2n} \sim a_1$, and a_{2n} is only adjacent to odd vertices, there are $(d(a_{2n}) - 1)$ odd values k such that (a_{2n}, a_k) is an edge and $f_{2m}(k)$ is defined, and a_1 is adjacent to $d(a_1) - 1$ even vertices other than a_{2n}. So the total number of values k such that either (a_{2n}, a_k) or $(a_1, a_{f_{2m}(k)})$ is an edge is $N = (d(a_{2n-1}) + d(a_1) - 1)$. If $d(a_1) + d(a_{2n}) \geq n + 1$, then $N \geq n - 1$. Either every possible value of k is represented at least once—so either (a_{2n}, a_k) or $(a_1, a_{f_{2m}(k)})$ is an edge for every k, and part (ii) is satisfied—or else both (a_{2n}, a_k) and $(a_1, a_{f_{2m}(k)})$ are edges, so there is a $2m$-cycle, contradicting the assumptions in part (ii). If $d(a_1) + d(a_{2n}) > n + 1$, then $N \geq n$, so there must be a value k such that both (a_{2n}, a_k) and $(a_1, a_{f_{2m}(k)})$ are edges, and there is a $2m$-cycle; this is true for every m, so G is bipancyclic, proving part (i). $\square$

Lemma 25.2 ([29]). *Suppose G is a bipartite graph containing a Hamilton cycle $C = (a_1, a_2, \ldots, a_{2n}, a_1)$.*

(i) *If $d(a_2) + d(a_{2n}) > n + 2$, then G is bipancyclic.*
(ii) *If $d(a_2) + d(a_{2n}) \geq n + 2$, and G is not bipancyclic, then both of the pairs (a_2, a_{2n-1}) and (a_{2n}, a_3) form edges in G and, for each odd integer k, $5 \leq k \leq 2n - 1$, exactly one of the pairs (a_{2n}, a_k) and $(a_2, a_{g_{2m}(k)})$ is an edge of G, where $g_{2m}(k) = 2n - 2m + k$ when $5 \leq k \leq 2m - 3$ and $g_{2m}(k) = k - 2m + 4$ when $2m - 1 \leq k \leq 2n - 1$.*

(In this case, the lemma applies if a_2 and a_{2n} are replaced by any two consecutive odd vertices or consecutive even vertices in C.)

Proof. The proof is almost identical to that of Lemma 25.1. Again we shall call a vertex a_k odd or even according as k is odd or even. Then both a_2 and a_{2n} are adjacent only to odd vertices.

If both (a_{2n}, a_k) and $(a_2, a_{g_{2m}(k)})$ are edges of G, then G contains a cycle of length $2m$: when $5 \leq k \leq 2m-3$, the cycle is $(a_2, a_3, \ldots, a_k, a_{2n}, a_{2n-1}, \ldots, a_{2n-2m+k}, a_2)$, while if $2m - 1 \leq k$, the cycle is $(a_1, a_2 a_{k-2m+4}, a_{k-2m+5}, \ldots, a_k, a_{2n}, a_1)$.

As k ranges through the odd integers from 5 to $2m - 3$, $g_{2m}(k)$ covers the odd integers from $2n - 2m + 5$ to $2n - 3$, and as k ranges through the odd integers from $2m - 1$ to $2n - 1$, $g_{2m}(k)$ covers the odd integers from 3 to $2n - 2m + 3$, and there are no repeats.

As $a_{2n} \sim a_1$ and possibly $a_{2n} \sim a_3$, and a_{2n} is only adjacent to odd vertices, there are at least $(d(a_{2n}) - 2)$ odd values k such that (a_{2n}, a_k) is an edge and $g_{2m}(k)$ is defined, and a_2 is adjacent to $d(a_2) - 2$ odd vertices other than a_1 and a_3.

So the total number of values k such that either (a_{2n}, a_k) or $(a_1, a_{f_{2m}(k)})$ is an edge is $N = (d(a_{2n}) - 2) + (d(a_2) - 2)$. If $d(a_2) + d(a_{2n}) \geq n + 2$, then $N \geq n - 2$. Either every possible value of k is represented at least once—so either (a_{2n}, a_k) or $(a_2, a_{g_{2m}(k)})$ is an edge for every k, and part (ii) is satisfied—or else both (a_{2n}, a_k) and $(a_1, a_{g_{2m}(k)})$ are edges, so there is a $2m$-cycle, contradicting the assumptions in part (ii). If $d(a_2) + d(a_{2n}) > n + 2$, then $N > n - 2$, so there must be a value k such that both (a_{2n}, a_k) and $(a_2, a_{g_{2m}(k)})$ are edges, and there is a $2m$-cycle; this is true for every m, so G is bipancyclic, proving part (i). $\square$

Lemma 25.3 ([29]). *Suppose n is odd, and G is a bipartite graph whose disjoint vertex sets $V_1 = \{x_1, x_2, \ldots, x_n\}$ and $V_2 = \{y_1, y_2, \ldots, y_n\}$ satisfy (6.1), (6.2), and (6.3). Either G is bipancyclic, or G contains $n + 3$ or more vertices with degree at least $\frac{1}{2}(n + 1)$ and every Hamilton cycle in G contains consecutive vertices x, y satisfying $d(x) = d(y) = \frac{1}{2}(n + 1)$.*

Proof. From Theorem 23, G is Hamiltonian. Select one such Hamilton cycle, C say, and assume there is a second labeling of the vertices so that $C = (a_1, a_2, \ldots, a_{2n}, a_1)$, where the even vertices constitute V_1 (in some order).

We shall define $p = \frac{1}{2}(n+1)$. If $d(x_p) \geq p+1$, then more than half the vertices in V_1 have degree at least $p+1$, so there must be two vertices in V_1 that are consecutive even vertices in the Hamilton cycle and the sum of their degrees is at least $(p+1) + (p + 1) = n + 3$. Without loss of generality, label them a_2 and a_{2n}. Then, by Lemma 25.2, G is bipancyclic.

Now suppose $d(x_p) \leq p$. Then, by (6.3), $d(y_{p-1}) \geq p$. But if $d(y_{p-1}) > p$, we would have two consecutive odd vertices in C with degree-sum greater than $n + 2$; if we change the second labeling so that the even vertices constitute V_2, the same argument shows that G is bipancyclic.

So, if G is not bipancyclic, we can assume that $d(y_{p-1}) = p$, and by symmetry $d(x_{p-1}) = p$. So G contains at least $n + 3$ vertices with degree at least p. So every Hamilton cycle in G must contain a pair of consecutive vertices x and y that both have degree at least p. If either degree was actually greater than p, then we have two consecutive vertices the sum of whose degrees is greater than $n + 1$, so G would be bipancyclic by Lemma 25.1. So, if G is not bipancyclic, G contains $n + 3$ or more vertices with degree at least $\frac{1}{2}(n + 1)$ and every Hamilton cycle in G contains consecutive vertices x, y satisfying $d(x) = d(y) = \frac{1}{2}(n + 1)$. $\square$

Lemma 25.4 ([29]). *Suppose n is even, and G is a bipartite graph containing a Hamilton cycle $C = (a_1, a_2, \ldots, a_{2n}, a_1)$ in which $d(a_2) = d(a_{2n}) = \frac{1}{2}n + 1$. Then G is bipancyclic.*

Proof. Let us assume G is missing a cycle of length $2m$. We have $d(a_2) = d(a_{2n}) = \frac{1}{2}n + 1$, so from Lemma 25.2 we know that (a_2, a_{2n-1}) and (a_{2n}, a_3) are edges of G. So G contains the 4-cycle $(a_2, a_3, a_{2n}, a_{2n-1}, a_2)$ and also the $(2n - 2)$-cycle that bypasses a_1. So $6 \leq 2m \leq 2n - 4$.

Also, Lemma 25.2 tells us that for any k, exactly one of (a_{2n}, a_k) and $(a_2, a_{g_{2m}(k)})$ is an edge of G. Consider $k = 2n - 3$, so that $g_{2m}(k) = 2n - 2m + 1$. If $(a_{2n-2m+1}, a_2)$ is an edge, then $(a_2, a_{2n-1}, a_{2n-2}, \ldots, a_{2n-2m+1}, a_2)$ would be a $2m$-cycle, so (a_{2n}, a_{2n-3}) must be an edge.

Further, if a_{2n} is adjacent to a_i but not to a_{i+2} for some $i \geq 2m - 3$, then $(a_2, a_{g_{2m}(i-2)}) = (a_2, a_{i-2m+6})$ is an edge of G, so G contains the $(2m)$-cycle $(a_2, a_{2n-1}, a_{2n-2}, a_{2n-3}, a_{2n}, a_i, a_{i-1}, \ldots, a_{i-2m+6}, a_2)$. So if (a_{2n}, a_i) is an edge, then $(a_{2n}, a_{i+2}), (a_{2n}, a_{i+4}), \ldots, (a_{2n}, a_{2n-3})$ are all edges in G.

On the other hand, if a_{2n} is adjacent to a_i but not to a_{i+2} for some $i \leq 2m - 5$, then $(a_2, a_{g_{2m}(i-2)}) = (a_2, a_{2n-2m+i-2})$ is an edge of G, so G contains the $(2m)$-cycle $(a_2, a_3, \ldots, a_i, a_{2n}, a_{2n-3}, a_{2n-4}, \ldots, a_{2n-2m+i-2}, a_2)$. So if (a_{2n}, a_i) is an edge, then $(a_{2n}, a_{i-2}), (a_{2n}, a_{i-4}), \ldots, (a_{2n}, a_3)$ are all edges in G.

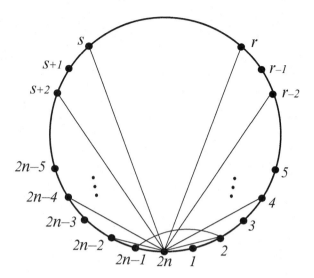

Fig. 6.1 Chords for Lemma 25.4

Now combine the above results. Under our assumption, the vertices adjacent to a_{2n} are $a_1, a_3, \ldots, a_r, a_s, a_{s+2}, \ldots, a_{2n-3}, a_{2n-1}$, where $r \leq 2m-5, s \geq 2m-3$, and $s - r = 2(n - d(a_{2n}) + 1)$. This is illustrated in Fig. 6.1 (for ease of reading, we use label i instead of a_i). The chord (a_2, a_{2n}) is also shown.

From the diagram, it is easy to see the cycle

$$C_1 = (a_{2n}, a_r, a_{r+1}, \ldots, a_s, a_{2n})$$

which has length $s - r + 2 = 2(n - d(a_{2n}) + 2)$. As $d(a_{2n}) = \frac{1}{2}n + 1$, C_1 has length $n + 2$. It is obvious from the diagram that G also contains cycles of every even length greater than that of C_1: for example, one can construct a cycle of length $n + 4$ by replacing the edge (a_{2n}, a_r) by the sequence $(a_{2n}, a_{r-2}, a_{r-1}, a_r)$ or by replacing (a_s, a_{2n}) by $(a_s, a_{s-1}, a_{s-2}, a_{2n})$, and so on. Also, the cycle

$$C_2 = (a_{2n}, a_r, a_{r-1}, \ldots, a_2, a_{2n-1}, a_{2n-2}, \ldots, a_s, a_{2n})$$

has length n, and similar modifications to those suggested for C_1 give cycles of all smaller even lengths down to 4. So G is bipancyclic. □

Lemma 25.5 ([29]). *Suppose n is even, and G is a bipartite graph with $2n$ vertices satisfying the conditions of Theorem 24. Then G is bipancyclic.*

Proof. Set $k = \frac{1}{2}n$. If $d(x_k) \geq k + 1$, then more than half the members of V_1 have degree at least $k + 1$. So in any Hamilton cycle there will be two consecutive members x_i and x_j of V_1 with degree at least $k + 1$, so the sum of the degrees is $d(x_i) + d(x_j) \geq 2k + 2 = n + 2$. If the sum is greater than $n + 2$, then G is bipancyclic by Lemma 25.2. Otherwise, $d(x_i) = d(x_j) = k + 1$, and by Lemma 25.4 G is bipancyclic.

On the other hand, if $d(x_k) \leq k$, then (6.3) implies $d(y_k) \geq k + 1$, and we can apply the same reasoning to V_2. So again G is bipancyclic.   □

This proves Theorem 25. We now need seven lemmas to prove:

Theorem 26. *Suppose G satisfies the conditions of Theorem 24 where n is odd. Suppose G contains at least $n + 3$ vertices with degree $\frac{1}{2}(n + 1)$ or greater, and a Hamilton cycle $C = (a_1, a_2, \ldots, a_{2n}, a_1)$ in which two consecutive vertices have degree exactly $\frac{1}{2}(n + 1)$. Then G is bipancyclic.*

From Lemma 25.1 it follows that if G is missing a $2m$-cycle, then for every k exactly one of (a_{2n}, a_k) and $(a_1, a_{f_{2m}(k)})$ is an edge of G; following [29] we shall refer to this as the $2m$-*principle*.

Lemma 26.1 ([29]). *If either (a_1, a_{2n-2}) or (a_{2n}, a_3) is an edge of G, then G is bipancyclic.*

Proof. Suppose (a_1, a_{2n-2}) is an edge of G but G contains no $2m$-cycle. The chord has deficit 2, so it generates a 4-cycle and a $(2n - 2)$-cycle, and therefore $6 \leq 2m \leq 2n - 4$.

If (a_1, a_{2n-2m}) were an edge of G then G would contain the $2m$-cycle $(a_1, a_{2n-2}, a_{2n-3}, \ldots, a_{2n-2m}, a_1)$, so (a_1, a_{2n-2m}) is not an edge, and by the $2m$-principle (a_{2n}, a_{2n-3}) is an edge.

Say (a_{2n}, a_i) is an edge in G. If (a_{2n}, a_{i+2}) were not an edge for some $i \geq 2m - 3$, then the $2m$-principle says (a_1, a_{i-2m+5}) is an edge, so G contains the $2m$-cycle

$$(a_1, a_{2n-2}, a_{2n-1}, a_{2n}, a_i, a_{i-1}, \ldots, a_{i-2m+5}, a_1).$$

So if (a_{2n}, a_i) is an edge of G for some $i \geq 2m - 3$, then so are (a_{2n}, a_{i+2}), $(a_{2n}, a_{i+4}), \ldots, (a_{2n}, a_{2n-3})$.

On the other hand, if (a_{2n}, a_i) is an edge but (a_{2n}, a_{i-2}) is not, for some $i \leq 2m-5$, then $(a_1, a_{2n-2m+i-1})$ is an edge by the $2m$-principle and we have a $2m$-cycle

$$(a_1, a_2, \ldots, a_i, a_{2n}, a_{2n-3}, a_{2n-4}, \ldots, a_{2n-2m+i-1}, a_1)$$

in G. So, if (a_{2n}, a_i) is an edge for some $i \leq 2m - 5$, then so are (a_{2n}, a_{i-2}), $(a_{2n}, a_{i-4}), \ldots, (a_{2n}, a_3)$.

So the vertices adjacent to a_{2n} are $a_1, a_3, \ldots, a_r, a_s, a_{s+2}, \ldots, a_{2n-1}$, where $r \leq 2m - 5, s \geq 2m - 3$, and $s - r = n + 1$.

Consider the cycles C_1 and C_2 that we defined in the proof of Lemma 25.4 (although now n is odd). Since C_1 has length $n + 3$, G contains a cycle of every even length $\geq n + 3$, and since C_2 has length $n - 1$, G contains a cycle of every even length $\leq n - 1$. So it must be that $2m = n + 1$.

To complete the proof of this lemma, we show that G contains a cycle of length $n + 1$. But if not, the $2m$-principle (with $2m = n + 1$) tells us that the vertices incident with a_1 are precisely $a_2, a_4, \ldots, a_{r+1}, a_{s+1}, a_{s+3}, \ldots, a_{2n-2}, a_{2n}$.

Now consider the $n - 1$ vertices a_i where $r + 1 < i < s$. G contains at least $n + 3$ vertices of degree $\frac{1}{2}(n + 1)$ or greater, so at least one of the $n - 1$ vertices a_i where $r + 1 < i < s$ must have degree at least $\frac{1}{2}(n + 1)$. So such a vertex a_i is adjacent to at least one vertex y, where $r \leq j \leq s + 1$ and $j \neq i \pm 1$. For convenience, assume $j < i$; as $j \neq i - 1$, we have $i - j \geq 3$. On the other hand, $i - j \leq (s - 1) - r = n - 2$. Select any two odd integers u and v, where $1 \leq u \leq j$ and $i \leq v \leq 2n - 1$. Then G contains the cycle

$$(a_{2n}, a_u, y, \ldots, a_i, a_{i+1}, \ldots, a_v, a_{2n})$$

which is of length $(v - u) + (j - i) + 3$. This cycle is of length at least 3 (case $u = j, v = i$) and at most $2n + 1 - (i - j) \geq n + 3$ (case $u = 1, v = 2n - 1$), and every even length in between $2n + 1 - (i - j)$ can be realized by a suitable choice of u and v. So length $n + 1$ will be achieved. The proof in the case $i < j$ is similar. □

In view of the above lemma, we can assume that G does not contain edges (a_1, a_{2n-2}) or (a_3, a_{2n}), so by the $2m$-principle both $(a_1, a_{2n-2m+4})$ and (a_{2m-3}, a_{2n}) are edges of G.

Lemma 26.2 ([29]). *If G is not bipancyclic, then there is an even integer i such that (a_1, a_i) is an edge of G, and so is either (a_{2n}, a_{i-1}) or (a_{2n}, a_{i+1}).*

Proof. Suppose not. Then the vertices adjacent to a_1 must be precisely $\{a_4, a_6, \ldots, a_{n-1}\}$ and those adjacent to a_{2n} must be precisely $\{a_{n+2}, a_{n+4}, \ldots, a_{2n-3}\}$; but in that case it is easy to check that G is bipancyclic. □

Observe that this lemma guarantees a 4-cycle, either $(a_1, a_i, a_{i+1}, a_{2n}, a_1)$ or $(a_1, a_i, a_{i-1}, a_{2n}, a_1)$.

Lemma 26.3 ([29]). *If G contains no $2m$-cycle, then $2m \geq n + 3$.*

Proof. Suppose $2m \leq n + 1$, and suppose the 4-cycle guaranteed by Lemma 26.2 is $(a_1, a_i, a_{i+1}, a_{2n}, a_1)$. Consider the two sets $S_1 = \{a_1, a_2, \ldots, a_{2n-2m+3}\}$ and $S_2 = \{a_{2m-2}, a_{2m-1}, \ldots, a_{2n}\}$. Since

$$2n - 2m + 3 \geq 2n - (n + 1) + 3 = n + 4 \geq n - 1 \geq 2m - 2,$$

S_1 and S_2 overlap, so both a_i and a_{i+1} belong to one of them. If it is S_1, then $(a_1, a_i, a_{i+1}, a_{2n}, a_{2n-1}, \ldots, a_{2n-2m+4}, a_1)$ is a $2m$-cycle in G, while if it is S_2 we have the $2m$-cycle $(a_1, a_i, a_{i+1}, a_{2n}, a_{2m-3}, a_{2m-4}, \ldots, a_2, a_1)$. If the 4-cycle was $(a_1, a_i, a_{i-1}, a_{2n}, a_1)$, replace a_{i+1} by a_{i-1} in the $2m$-cycles and we obtain the same result. In either case we have a contradiction, so the assumption that $2m \leq n + 1$ must be false. □

Lemma 26.4 ([29]). *If G contains no $2m$-cycle, then none of (a_1, a_{2m}), $(a_1, a_{2m+2}), \ldots, (a_1, a_{2n-2})$ is an edge of G.*

Proof. We know from Lemma 26.1 that (a_1, a_{2n-2}) is not an edge. If (a_1, a_{2m}) were one, then G would contain the $2m$-cycle $(a_1, a_2, \ldots, a_{2m}, a_1)$. If $2m \geq 2n - 4$, we are finished. So assume $2m + 2 \leq 2n - 4$, and assume (a_1, a_j) is an edge for some j, $2m + 2 \leq j \leq 2n - 4$.

We now show that there is an i, $2m + 2 \leq i \leq 2n - 4$, such that (a_1, a_i) is an edge of G but (a_1, a_{i+2}) is not. For, otherwise, (a_1, a_{2n-4}) is not an edge (or else we would have case $i = 2n - 4$), so (a_1, a_{2n-6}) is not an edge (or else case $i = 2n - 6$), and so on; but this process must stop sometime, because our assumption tells us that (a_1, a_j) is an edge for some $j \geq 2m + 2$.

So assume i is such that (a_1, a_i) is an edge of G but (a_1, a_{i+2}) is not, where $2m + 2 \leq i \leq 2n - 4$. Then $(a_{2n}, a_{2m-2n+i+1})$ is an edge by the $2m$-principle. As (a_1, a_{2m}) is not an edge, the $2m$-principle also implies that $(a_{2n}, a_{4m-2n-1})$ is also an edge. So G contains the $2m$-cycle $(a_1, a_2, \ldots, a_{4m-2n-1}, a_{2n}, a_{2m-2n+i+1}, \ldots, a_{i-1}, a_i, a_1)$, a contradiction. $\qquad\square$

Lemma 26.5 ([29]). *Assume G contains no $2m$-cycle. If G contains the edge (a_{2n}, a_i) for some $i \geq 2m + 1$, then each of $(a_{2n}, a_i), (a_{2n}, a_{i+2}), \ldots, (a_{2n}, a_{2n-1})$ is an edge of G.*

Proof. Assume (a_{2n}, a_i) is an edge of G but (a_{2n}, a_{i+2k}) is not, for some k satisfying $2 \leq 2k \leq 2n - i - 1$. Write G' for the graph formed by deleting vertices $\{a_{i+1}, a_{i+2}, \ldots, a_{2n-1}\}$ from G. Then G' contains the Hamilton cycle $(a_1, a_2, \ldots, a_i, a_{2n}, a_1)$. As a_{2n} is adjacent to at most $\frac{1}{2}(2n - i - 3)$ of the deleted vertices, and (by Lemma 26.4) a_1 is adjacent to none of them, and since $|V(G')| = i + 1$, therefore $d_{G'}(a_1) + d_{G'}(a_{2n}) \geq (n + 1) - \frac{1}{2}(2n - 1 - 3) = \frac{1}{2}(i + 1) - 2 > \frac{1}{2}(|V(G')| + 1)$. So, from Lemma 25.1 G' is bipancyclic. But $|V(G')| = i + 1 > 2m$, so G' contains a $2m$-cycle, whence G does also. Therefore each of the named edges must lie in G. $\qquad\square$

Lemma 26.6 ([29]). *If G is missing a $2m$-cycle, then $2m = 2n - 2$.*

Proof. $(a_1, a_{2n-2m+4})$ is an edge of G but (a_1, a_{2m}) is not, so it follows that there is some value i with $2n - 2m + 4 \leq i \leq 2m - 2$ such that (a_1, a_i) is an edge of G but (a_1, a_{i+2}) is not. But, if (a_1, a_{i+2}) is not an edge, the $2m$-principle ensures that $(a_{2n}, a_{2m-2n+i+1})$ is an edge, where $2m - 2n + i + 1 \geq 5$.

If (a_{2n}, a_{2n-3}) is an edge of G, then G contains the $2m$-cycle
$(a_1, a_2, \ldots, a_{2m-2n+i+1}, a_{2n}, a_{2n-3}, a_{2n-4}, \ldots, a_i, a_1)$. On the other hand, if there is no edge (a_{2n}, a_{2n-3}) and $2m < 2n - 2$, then Lemma 26.5 shows that (a_{2n}, a_{2m+1}) is not an edge, so by the $2m$-principle (a_1, a_4) is an edge. But again we have a $2m$-cycle, namely $(a_1, a_4, a_5, \ldots, a_{2m-2n+i+1}, a_{2n}, a_{2n-1}, \ldots, a_i, a_1)$. $\qquad\square$

Lemma 26.7 ([29]). *If G contains no $(2n-2)$-cycle, then vertices $a_{2n-2}, a_{2n-1}, a_2,$ and a_3 all have degree less than $\frac{1}{2}(n + 1)$.*

Proof. From Lemma 26.1, neither (a_1, a_{2n-2}) nor (a_{2n}, a_3) is an edge of the non-bipancyclic graph G. If $2m = 2n - 2$, the $2m$-principle shows that (a_1, a_6) and (a_{2n}, a_{2n-5}) are both edges of G (see Fig. 6.2). Edges

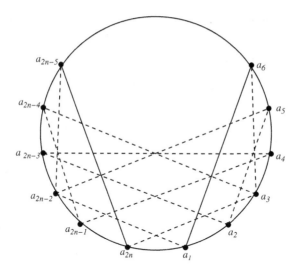

Fig. 6.2 *Dash lines* represent missing edges in the graph

$(a_2, a_5), (a_3, a_6), (a_3, a_{2n}), (a_{2n-1}, a_{2n-4}), (a_{2n-2}, a_{2n-5})$ would have deficit 3, so they would yield a $(2n-2)$-cycle. Similarly, the edges (a_3, a_{2n-4}), (a_4, a_{2n-1}), (a_4, a_{2n-3}) and (a_5, a_{2n-2}) lead to the $(2n-2)$-cycles

$$(a_{2n-4}, a_{2n-3}, \ldots, a_{2n}, a_{2n-5}, a_{2n-6}, a_{2n-7}, \ldots, a_3, a_{2n-4}),$$
$$(a_1, a_2, a_3, a_4, a_{2n-1}, a_{2n-2}, \ldots, a_6, a_1)$$
$$(a_1, a_2, a_3, a_4, a_{2n-3}, a_{2n-2}, a_{2n-1}, a_{2n}, a_{2n-5}, a_{2n-6}, \ldots, a_6, a_1) \text{ and}$$

$(a_1, a_2, a_3, a_4, a_5, a_{2n-2}, a_{2n-3}, \ldots, a_6, a_1)$, respectively. Hence, these edges are not in G.

Define a set of vertices $S = \{a_1, a_2, \ldots, a_5, a_{2n-4}, a_{2n-3}, \ldots, a_{2n}\}$. As we said, (a_1, a_{2n-2}) is not an edge; similarly (a_1, a_4) is not an edge (or we could obtain a $(2n-2)$-cycle by omitting a_2, a_3 from C). So a_1 is adjacent only to a_2, a_{2n}, and possibly a_{2n-4} in S. So a_1 is adjacent to at least $\frac{1}{2}(n+1) - 3$ members of $V(G) \setminus S$, with equality if and only if (a_1, a_{2n-4}) is an edge.

Say (a_1, a_i) is an edge, where $a_i \in V(G) \setminus S$. Then (a_{2n-1}, a_{i-2}) is not an edge (or else we would have the $(2n-2)$-cycle $(a_1, a_2, \ldots, a_{i-2}, a_{2n-1}, a_{2n-2}, \ldots, a_1)$, but this is impossible. So a_{2n-1} is non-adjacent to at least $\frac{1}{2}(n+1) - 3$ vertices, namely a_{i-2} where $4 \leq i - 2 \leq 2n - 8$. Moreover, a_{2n-2} is not adjacent to a_1 or a_{2n-5}. So a_{2n-1} has degree at most $n - [\frac{1}{2}(n+1) - 3] - 2$, that is $\frac{1}{2}(n+1)$, with equality only if (a_1, a_{2n-4}) is an edge.

But if a_{2n-1} has degree $\frac{1}{2}(n+1)$, the same argument we used to show that (a_{2n}, a_{2n-5}) is an edge would show that (a_{2n-1}, a_{2n-6}) is an edge, so we have the $(2n-2)$-cycle $(a_1, a_2, \ldots, a_{2n-6}, a_{2n-1}, a_{2n-2}, a_{2n-3}, a_{2n-4}, a_1)$. So $d(a_{2n-1}) < \frac{1}{2}(n+1)$. Similarly $d(a_2) < \frac{1}{2}(n+1)$.

Again, if (a_1, a_i) is an edge, where $a_i \in V(G) \setminus S$, then (a_{2n-2}, a_{i-1}) is not an edge (or else we would have the $(2n - 2)$-cycle $(a_1, a_2, \ldots, a_{i-1}, a_{2n-3}, a_{2n-2}, \ldots, a_1)$, again impossible. So a_{2n-2} is non-adjacent to at least $\frac{1}{2}(n + 1) - 3$ vertices, namely a_{i-1} where $5 \leq i - 1 \leq 2n - 7$. Moreover, a_{2n-2} is not adjacent to a_1 or a_{2n-5}. So a_{2n-2} also has degree at most $\frac{1}{2}(n + 1)$, again with equality only if (a_1, a_{2n-4}) is an edge.

If (a_{2n-2}, a_3) is an edge, then G would contain the $(2n - 2)$-cycle $(a_1, a_{2n}, a_{2n-1}, a_{2n-2}, a_3, a_4, \ldots, a_{2n-4}, a_1)$. So (a_{2n-2}, a_3) is not an edge, and $d(a_{2n-2}) < \frac{1}{2}(n + 1)$. In the same way, it can be seen that $d(a_3) < \frac{1}{2}(n + 1)$. □

Lemma 26.8 ([29]). *G contains a $(2n - 2)$-cycle.*

Proof. In the preceding lemmas, vertices a_{2n} and a_1 could be replaced by any two consecutive vertices in C with degree at least $\frac{1}{2}(n + 1)$. So it would follow that at most half the vertices in G have degree at least $\frac{1}{2}(n + 1)$. But this contradicts the fact that at least $n + 3$ vertices have degree at least $\frac{1}{2}(n + 1)$. So the assumption that G contains no $(2n - 2)$-cycle cannot be true. □

Lemmas 26.6 and 26.8 prove the theorem.

Chapter 7
Uniquely Bipancyclic Graphs

7.1 Introduction

In this chapter we look at the problem of uniquely bipancyclic graphs, that is bipartite graphs that contain exactly one cycle of each length from 4 up to the number of vertices. If such a graph contains c cycles, their lengths are 4, 6, ..., $2c + 2$, so the graph has $2c + 2$ vertices; moreover the sum of the lengths of the cycles (total number of edges in the cycles, with multiple appearances in different cycles counted multiply) is $4 + 6 + \cdots + (2c + 2) = c^2 + 3c$. We shall denote this by $s(c)$.

Suppose we have a bipancyclic graph on $2n$ vertices. It must contain a Hamilton cycle, so the graph could be represented as a cycle of length $2n$ together with some other edges which we call *chords*, just as we did in the pancyclic case.

In diagrams, we represent our graph as a circle with the chords as straight lines. The segments of the outer circle may contain a number of vertices, but the chords only have vertices at their ends. If vertex labels are needed, we assume the vertices are $a_1, a_2, \ldots, a_{2n}$ in clockwise order. Several of the diagrams were used earlier, but we repeat them for convenience.

As we pointed out when considering minimal pancyclicity, if a $2n$-vertex graph contains a chord (a_x, a_y), there will be two cycles containing it, one obtained by going from a_x to a_y clockwise and then along the edge (a_y, a_x), the other by going anticlockwise. The numbers of vertices in these two cycles will total $2n + 2$, because the edges on the cycle will be counted once each and the chord counted twice. We shall say a chord is *of type* (p, q) if the two cycles it generates have lengths p and q. If there are two chords, there may be one or two cycles that contain both of them; in the latter case, the total is $2n + 4$. Any collection of chords generates either one or two cycles that contain them all; if there are c chords and together they generate two cycles, then the two lengths will add to $v + 2c$.

© The Author(s) 2016
J.C. George et al., *Pancyclic and Bipancyclic Graphs*, SpringerBriefs
in Mathematics, DOI 10.1007/978-3-319-31951-3_7

In the next three sections we look at the cases that can be solved by hand; after that, we look at computer methods. The results in this chapter come from [21, 26, 35].

7.2   Graphs with Fewer than Two Chords

If there are no chords, the graph contains only one (Hamilton) cycle, of length $2n$, and the only bipancyclic case is $v = 4$. If there is one chord, there are two further (one-chord) cycles. So, if the graph has only one chord, there are exactly three cycles. These cycles were illustrated in Fig. 4.1, and are reproduced in Fig. 7.1 (there are three drawings of the same graph, with the cycles shown in bold).

Fig. 7.1 Cycles in the case of one chord

If the graph is uniquely bipancyclic the lengths of the cycles must be 4, 6, and 8. This can be achieved by inserting a chord a_1a_4 into a cycle of length 8; any other example will obviously be isomorphic to this.

One can construct a bipancyclic graph on six vertices, again by inserting chord a_1a_4, but the graph will contain two cycles $(a_1, a_2, a_3, a_4, a_1)$ and $(a_1, a_6, a_5, a_4, a_1)$ of length 4, so it is not uniquely bipancyclic (see Fig. 7.2).

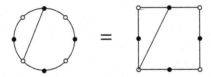

Fig. 7.2 The uniquely bipancyclic graph of order 8

As we shall see in the next section, any uniquely bipancyclic graph with more than one chord will have at least six cycles and therefore at least 14 vertices, so there are no uniquely bipancyclic graphs on 6, 10, or 12 vertices.

7.3   Two Chords

There are three possible patterns for two chords: case A, where the chords share an endpoint; case B, where they do not cross in the standard diagram, and case C, where they cross. The three types were illustrated in Fig. 4.2 and appear again in Fig. 7.3.

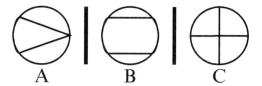

Fig. 7.3 The possible cases with two chords

In addition to the cycles that contain no chord or one chord, there will be new cycles that contain two chords. Let us count cycles in the three cases:

A. There is one new cycle, so together with the Hamilton cycle and the four one-chord cycles (two per chord) there are six cycles in total;
B. There is one new cycle, for six in total;
C. There are two new cycles, for seven in total.

Types A and B produce 6 cycles (including the Hamilton cycle) so a uniquely bipancyclic graph of type A or B would have 14 vertices; one of type C would have 16.

In the 14-vertex case, the total number of edges in the two cycles containing a given chord (but not the other one) will total 16, so four cycles each containing exactly one chord will contain a total of 32 edges, and the Hamilton cycle contains 14, so if the cycle containing both chords has z edges, then

$$14 + 32 + z = 4 + 6 + 8 + 10 + 12 + 14 = 54$$

and $z = 8$.

In the 16-vertex case, the total number of edges in the two cycles containing a given chord (but not the other one) will total 18, the two cycles that contain both chords will have a total of 20 edges, and the Hamilton cycle contains 16. So the seven cycles have a total of $16 + 18 + 18 + 20 = 72$ edges. But $4 + 6 + 8 + 10 + 12 + 14 + 16 = 70$. So there is no uniquely bipancyclic graph on 16 vertices.

Type A

The four cycles containing one edge each must contain 4, 6, 10, and 12 edges. So one chord must be of type (4,12) and the other of type (6,10). Up to isomorphism, we can assume the common endpoint of the chords to be a_1 and the first chord to be (a_1, a_4). The second must be (a_1, a_6) or (a_1, a_{10}). Only the second choice causes the cycle containing both chords to be of length 8. So there is exactly one solution.

Without loss of generality the chords are (a_1, a_i) and (a_j, a_k) where, in order for the cycle containing both chords to have length 8, $(j - i) + (15 - k) = 6$. (The eight edges include the two chords.) The first chord is of type $(i, 16 - i)$ and the second is of type $(k - j + 1, 15 - j + k)$. One of these pairs must be (4,12), so let us take $i = 4$. Then $(j - i) + (15 - k) = 6$ becomes $j + 11 - k = 6$ or $k = j + 5$. Then type $(k - j + 1, 15 - j + k)$ becomes type (6,10), as required. There are five possibilities:

$j = 5, 6, 7, 8$, or 9. Cases $j = 5$ and $j = 9$ are easily seen to be isomorphic, as are cases $j = 6$ and $j = 8$. So there are three solutions, giving a total of four non-isomorphic uniquely bipancyclic graphs of order 14.

The four uniquely bipancyclic graphs are shown in Fig. 7.4.

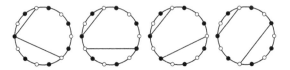

Fig. 7.4 The possible cases with two chords

7.4 Three Chords

If a graph has three chords, we classify by looking at the three pairs of chords. We refer to the configuration by the string of three letters corresponding to the three types of chord interaction. For example, type AAB is a graph in which two of the pairs of chords are type A (they do not cross, and have no common endpoint) and one pair is type B (they have a common endpoint). There are 14 types of graph: AAAi, AAAii, AABi, AABii, AAC, ABBi, ABBii, ABC, ACC, BBBi, BBBii, BBC, BCC, and CCC. (There are two types AAA, two types AAB, and two types ABB (one where the three chords form a "C" pattern and one where they form a "Z"), and two types BBB (one where all three chords have a common endpoint and one where they form a triangle).) They are illustrated in Fig. 7.5, below, which is the same as Fig. 4.4 in Chap. 4.

The following table, also from Chap. 4, counts the number c of cycles in a graph, in each of the 14 cases. $C(n)$ means the number of cycles involving exactly n chords. We have added rows showing the number n of vertices required for a uniquely bipancyclic graph with this chord structure, and the total $s(c)$ of the edges in the cycles.

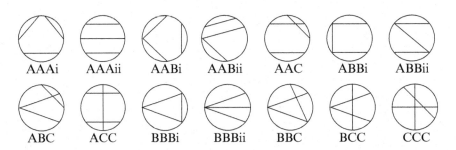

Fig. 7.5 Cases of three chords

	AAAi	AAAii	AABi	AABii	AAC	ABBi	ABBii
C(0)	1	1	1	1	1	1	1
C(1)	6	6	6	6	6	6	6
C(2)	3	3	3	3	4	3	3
C(3)	1	0	1	0	1	1	0
c	11	10	11	10	12	11	10
n	24	22	24	22	26	24	22
$s(c)$	154	130	154	130	180	154	130

	ABC	ACC	BBBi	BBBii	BBC	BCC	CCC
C(0)	1	1	1	1	1	1	1
C(1)	6	6	6	6	6	6	6
C(2)	4	5	3	3	4	5	6
C(3)	1	2	1	0	1	1	2
c	12	14	11	10	12	13	15
n	26	30	24	22	26	28	32
$s(c)$	180	238	154	130	180	208	270

Whenever a diagram is shown in the discussion of a chord pattern, lower-case letters in the figure are the numbers of edges between the points where chord and basic cycle meet; we refer to these sets of edges as *arcs* and the number of edges is the *arc length*. Capitals are chord names. We identify cycles by the chords that they contain: for example, "an XY cycle" will mean one that contains chords X and Y and no others.

Given a chord X, the lengths of the two X cycles will add to $2n + 2$, because they each contain the edge X and every edge of the Hamilton cycle appears once. Similarly, if chords X and Y are of type C, the lengths of the two XY cycles add to $2n + 4$. If there are two cycles containing chords X, Y, and Z, their lengths total $2n + 6$.

Type AAAi

This graph has $n = 24$ vertices. There will need to be 11 cycles, totalling 154 edges. The Hamilton cycle and the six one-chord cycles have a total of $n + 3(n + 2) = 102$ edges. Therefore the remaining three cycles have a total of 52 edges (see Fig. 7.6).

The cycle containing both chords X and Y has $b + d + e + f + 2$ edges, the YZ cycle has $a + b + d + f + 2$ edges, the XZ cycle has $b + c + d + f + 2$ edges, and the XYZ cycle has $b + d + f + 3$ edges. So the four cycles together total $3(b + d + f) + (a + b + c + d + e + f) + 9$ edges. Since $a + b + c + d + e + f = 24$, we have $3(b + d + f) = 19$, so $b + d + f$ is not an integer, a contradiction. So there is no uniquely bipancyclic graph with chord pattern AAAi.

Type AAAii

This graph has $n = 22$ vertices. There will need to be 10 cycles, totalling 130 edges. The Hamilton cycle and the six one-chord cycles have a total of $n + 3(n + 2) = 94$ edges. Therefore the remaining three cycles have a total of 36 edges (see Fig. 7.7).

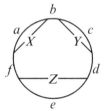

Fig. 7.6 Chord type AAAi

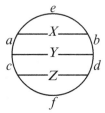

Fig. 7.7 Chord type AAAii

The cycle containing both chords X and Y has $a + b + 2$ edges, the YZ cycle has $c + d + 2$ edges, and the XZ cycle has $a + b + c + d + 2$ edges. So the three cycles together total $2(a + b + c + d) + 6$ edges. This equals 36, so $a + b + c + d = 15$, and $e + f = 7$. But there are cycles of lengths $e + 1$ and $f + 1$, so e and f must both be odd, which is a contradiction. So there is no uniquely bipancyclic graph—and in fact no bipancyclic graph—with chord pattern AAAii.

Type AABi
The graph will have $n = 24$ vertices; $a + b + c + d + e = 24$. There are 11 cycles, of lengths $24, 1 + a, 1 + b, 1 + e, 25 - a, 25 - b, 25 - e, c + d + e + 2, b + c + d + 2, a + c + d + 2$, and $c + d + 3$, and for a uniquely bipancyclic graph these must all be different. As the cycles have even length a, b, e are all odd numbers and $c + d$ is also odd. Without loss of generality we can assume $a < b < e$. Then the only way to have a cycle of length 4 is if $a = 3$. To form a 6-cycle, we must have $d + e = 3$ or $b = 5$. In the former case, we must have $b + 1 \geq 8$, so $b \geq 7$. If $b = 7$, then $b + 1 = 8 = a + c + d$, a repeated cycle length. If $b > 7$ we would have $e \leq b$, a contradiction. So $d + e = 3$ is impossible, and $b = 5$. There are three cases: $e = 7, c + d = 9$, and $25 - e = 18 = c + d + e + 2$; $e = 9, c + d = 7$, and $e + 1 = 10 = a + c + d$; or $e = 11, c + d = 5$, and $e + 1 = 12 = b + c + d + 2$. Therefore there is no uniquely bipancyclic graph with chord pattern AABi (see Fig. 7.8).

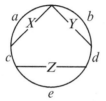

Fig. 7.8 Chord type AABi

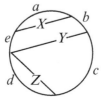

Fig. 7.9 Chord type AABii

Type AABii

This pattern involves 10 cycles (including the Hamilton cycle). So, if there is a uniquely bipancyclic graph with this pattern, it has cycles of lengths 4, 6, 8, 10, 12, 14, 16, 18, 20, and 22. So the number of vertices is 22, and $a + b + c + d + e = 22$. Each of a, b, c, d, e is greater than 0 (see Fig. 7.9).

Here are the cycle lengths (left column names them; second column shows all chords in the cycle.)

C1	—	22
C2	X	$a + 1$
C3	X	$b + c + d + e + 1 = 21 - a$
C4	Y	$c + d + 1$
C5	Y	$a + b + e + 1 \quad = 21 - c - d$
C6	Z	$d + 1$
C7	Z	$a + b + c + e + 1 = 21 - d$
C8	XY	$b + e + 2 \qquad = 24 - a - c - d$
C9	XZ	$b + c + e + 2 \quad = 24 - a - d$
C10	YZ	$c + 2.$

Since all cycles must be even, we must have

a is odd (from C2)

d is odd (from C6)

$c + d$ is odd (from C4) so c is even (also follows from C10)

$b + e$ is even (from C8).

Also, since all cycles are length 4 or greater,

$$a \geq 3 \text{ (from } C2)$$
$$d \geq 3 \text{ (from } C6)$$

and since no two cycles are the same length,

$$a \neq d \text{ (from } C2 \text{ and } C6).$$

Let us assume $a > d$. It then follows that $a > 3$.

Now there must be a cycle of length 4. Candidates are

$C2$, implying $a = 3$, which is impossible

$C6$, implying $d = 3$

$C8$, implying $b = e = 1$

$C10$, implying $c = 2$.

Three other cases,

$C2$, implying $a = 3$

$C4$, implying $c = 2, d = 1$

$C5$, implying $a = b = d = 1$

are ruled out, the first by our assumption and the other two by the fact that d is each at least 3.

C6: This does not work, because both $C3$ and $C9$ would be length $21 - a$.

C8: We assume $b = e = 1$. Also $a \geq 5$ and $d \geq 5$ (both are odd, ≥ 3, and if either equaled 3 we have two cycles length 4). Similarly $c \geq 4$. Cycle lengths are

$C1 : 22, \quad C2 : a + 1, \quad C3 : 21 - a, \ C4 : c + d + 1, \ C5 : 21 - c - d$
$C6 : d + 1, \ C7 : 21 - d, \ C8 : 4, \quad\quad C9 : c + 4, \quad\quad C10 : c + 2.$

As $a + c + d = 20$, we can get upper bounds on a, c, and d by using the given lower bounds. For example, $d \geq 5$ and $c \geq 4$ imply $a \leq 11$, so $5 \leq a \leq 11$; similarly $5 \leq d \leq 11$ and $4 \leq c \leq 10$. (Actually, as $a \neq d$, $a + d \geq 12$, and $c \leq 8$.)

So how do we get a cycle of length 20? The only case not immediately eliminated is $C4$, but that would need $c + d = 19$, so $a = 1 \ldots$ impossible.

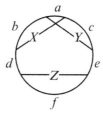

Fig. 7.10 Chord type AAC

C10: We assume $c = 2$. Again $a \geq 5$ and $d \geq 5$ (both are odd, ≥ 3, and if either equaled 3 we have two cycles length 4). Also $b + e \geq 4$. Cycle lengths are

$$C1 : 22, \quad C2 : a + 1, \quad C3 : 21 - a \quad C4 : d + 3, \quad C5 : 19 - d$$
$$C6 : d + 1, \, C7 : 21 - d, \, C8 : 22 - a - d, \, C9 : 24 - a - d, \, C10 : 4.$$

The only way to get a cycle of length 20 is if $d = 17$ or 19. But $a + b + c + e \geq 9$, so $d \leq 13$. So there is no uniquely bipancyclic graph with chord pattern AABii.

Type AAC

A uniquely bipancyclic graph will contain 12 cycles, $n = 26$ and the chords total 180 edges. The cycles with 0 and 1 chord and two XY cycles total 140 edges; XZ has $2 + b + c + e$, YZ has $2 + b + d + e$, and XYZ has $3 + a + d + e$. So $7 + a + b + c + 3d + 3e = 40$. Using the fact that $a + b + c + d + e + f = 26$, we have $33 + 2(d + e) - f = 40$, or $2(d + e) - f = 7$. Therefore $2(d + e + f) = 7 + 3f$, and $2(a + b + c) = 2(a + b + c + d + e + f) - 2(d + e + f) = 52 - 7 - 3f = 45 - 3f$. Up to isomorphism we can assume $b \geq c$ (see Fig. 7.10).

The cycle lengths are

$$a + b + 1 \quad 27 - a - b \quad a + c + 1 \quad 27 - a - c$$
$$f + 1 \quad 27 - f \quad 2 + b + c \quad 28 - b - c$$
$$2 + c + d + e \quad 2 + b + d + e \quad 3 + a + d + e \quad 26.$$

So b and c cannot be equal, or there would be two equal cycles. Also b and c are of the same parity, opposite to that of a. Moreover, none of the 1-chord cycle lengths $a + b + 1, a + c + 1$, and $f + 1$ can equal 14, as the other cycle associated with the same chord would also be of length 14.

The only possible lengths for a 4-cycle are $a + b + 1, a + c + 1, f + 1$, and $2 + b + c$. But $a + b + 1 = 4$ implies $a + c + 1 < 4$, which is impossible, and $2 + b + c = 4$ implies $b = c = 1$, also impossible.

Suppose $a + c + 1 = 4$. Then $2(a + b + c) = 45 - 3f$ becomes $2b + 6 = 45 - 3f$ or $2b = 39 - 3f$. So b is a multiple of 3. If $b = 15$, then $f = 3$, which gives another 4-cycle. So $b = 3, 6, 9,$ or 12. Clearly $\{a, c\} = \{1, 2\}$, but if $c = 1$, then $a + b + 1 = 2 + b + c$, so $c = 2, a = 1$, and therefore b is even. There remain two cases:

$b = 6, a = 1, c = 2, f = 9, d + e = 8$, in which $f + 1 = 2 + b + c = 10$;

$b = 12, a = 1, c = 2, f = 5, d + e = 6$, in which $a + b + 1 = 27 - a - b = 14$.

So there are no examples with $a + c + 1 = 4$.

Suppose $f + 1 = 4$. Then $d + e = 5$ and $a + b + c = 18$. a must be even and b and c odd. If we avoid cases where $a + b + 1$ or $a + c + 1$ equals 14, there remain ten cases, all with $f = 3$ and $d + e = 5$:

$a = 2, b = 9, c = 7$, in which $27 - a - c = 2 + b + c = 18$;

$a = 2, b = 13, c = 3$, in which $27 - a - b = 28 - b - c = 12$;

$a = 4, b = 11, c = 3$, in which $a + b + 1 = 2 + b + c = 16$;

$a = 4, b = 13, c = 1$, and no problem arises;

$a = 6, b = 9, c = 3$, in which $a + b + 1 = 28 - b - c = 16$;

$a = 6, b = 11, c = 1$, in which $a + c + 1 = 2 + c + d + e = 8$;

$a = 8, b = 7, c = 3$, in which $27 - a - b = 1 + a + c = 12$;

$a = 8, b = 9, c = 1$, in which $27 - a - b = 1 + a + c = 10$;

$a = 10, b = 7, c = 1$, in which $27 - a - b = 2 + b + c = 10$;

$a = 14, b = 3, c = 1$, in which $27 - f = 28 - b - c = 24$.

So there is one solution. This gives rise to four uniquely bipancyclic graphs with chord pattern AAC, according as $(d, e) = (1, 4), (2, 3), (3, 2)$, or $(4, 1)$. An example is shown in Fig. 7.11.

Type ABBi
We have $n = 24$ and the chords total 154 edges. A uniquely bipancyclic graph will contain 11 cycles. The Hamilton cycle and the six one-chord cycles have a total of $n + 3(n + 2) = 102$ edges. Therefore the remaining four cycles have a total of 52 edges. The XY, XZ, YZ, and XYZ cycles have $(b + c + 2), (b + d + 2), (a + b + 2)$, and $(b + 3)$ edges, respectively, for a total of $(a + b + c + d) + 3b + 9$, which equals $3b + 33$ as $a + b + c + d = n$. So $3b = 19$, which is not a multiple of 3—a

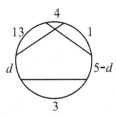

Fig. 7.11 Uniquely bipancyclic graphs of type AAC: $d = 1,2,3,4$

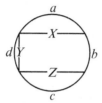

Fig. 7.12 Chord type ABBi

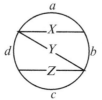

Fig. 7.13 Chord type ABBii

Fig. 7.14 Chord type ABC

contradiction. So there is no uniquely bipancyclic graph with chord pattern ABBi (see Fig. 7.12).

Type ABBii
In this case $n = 22$ and the chords total 130 edges. A uniquely bipancyclic graph will contain 10 cycles. The Hamilton cycle and the six one-chord cycles have a total of 94 edges. Therefore the remaining four cycles have a total of 36 edges. The XY, XZ, and YZ cycles have $(b + 2)$, $(b + d + 2)$, and $(d + 2)$ edges, respectively, for a total of $2(b + d) + 6$, so $b + d = 15$, and the XZ cycle has 17 vertices, an odd number. So there is no uniquely bipancyclic graph with chord pattern ABBii (see Fig. 7.13).

Type ABC
A uniquely bipancyclic graph will contain 12 cycles, $n = 26$ and the chords total 180 edges. The cycles with 0 and 1 chord and two XY cycles total 140 edges; the XZ cycle has $2+c+e$, YZ's has $2+b+c$, and XYZ's has $3+a+c$. So $7+a+b+3c+e = 40$. Using the fact that $a + b + c + d + e = 26$, we have $33 + 2c - d = 40$, or $2c - d = 7$ (see Fig. 7.14).

The cycle lengths are (Hamilton) 26; (X): $a+b+1, c+d+e+1$; (Y): $a+e+1$, $b+c+d+1$; (Z): $d+1, a+b+c+e+1$; (XY): $a+c+d+2, b+e+2$; (XZ): $c+e+2$; (YZ): $b+c+2$; (XYZ): $a+c+3$. From these we see that $a \neq b+1$ (or the cycle lengths $a+e+1$ and $b+e+2$ would be equal), $a \neq e+1$ (or the cycle lengths $a+b+1$ and $b+e+2$ would be equal), and $d \neq 1$ (or there would be a cycle of length 2); since d is odd, $d \geq 3$ and therefore $c \geq 5$. The only possible cycles of length 4 have lengths $a+b+1$ (which would imply $a=1, b=2$), $a+e+1$ (which would imply $a=1, e=2$), $b+e+2$ (which would imply $b=e=1$), and $d+1$ (so $d=3$).

Case $a=1, b=2$: then $c+d+e=23$; from $2c-d=7$, we get $3c+e=30$. So e is a multiple of 3. As $a+e+1 = e+2$ is even, e is even, so $e=6, 12, 18$, or 24. But $c+d \geq 8$, so the only possibilities are $e=6$ and $e=12$. If $e=6$, then $c=8$, so $d=9$, so $d+1=10=b+e+2$, and there are two cycles of length 10. If $e=12$, then $c=6, d=5$, and $a+e+1=14=b+c+d+1$, and there are two cycles of length 14. Neither case is uniquely bipancyclic.

Case $a=1, e=2$: $b+c+d=23$, and similarly to the above we deduce $b+3c=30$, b is even, so 6 divides b. The only possibilities are $b=6, c=8, d=9$, whence $d+1=10=b+e+2$, and there are two cycles of length 10, and $b=12, c=6, d=5$, so $a+b+1=14=c+d+e+1$, and there are two cycles of length 14. Neither case is uniquely bipancyclic.

Case $b=e=1$ is impossible because the cycles of length $a+b+1$ and $a+e+1$ would be of the same length.

Case $d=3$: $c=5$. Clearly a is even and b and e are odd. Cycle lengths are $26, a+b+1, e+9, a+e+1, b+9, 4, 24$ (and $a+b+e = 18$), $a+10, b+e+2, e+7, b+7, a+8$. Clearly b and e differ by at least 4 (otherwise $\{e+7, b+7, e+9, b+9\}$ must contain a duplication) so $b+e \geq 6$ and therefore $a \leq 12$. If $a=2$ we have cycles of length 10 and 12, so neither b nor e can equal 3 or 5; the only solution for $b+e=16$ is that they equal 7 and 9, so they do not differ by at least 4, a contradiction. If $a=4$ the only possibilities for a cycle of length 6 are the $a+b+1$ and $a+e+1$ cycles, with $b=1$ and $e=1$, respectively. Both of these work, yielding two uniquely bipancyclic graphs on 26 vertices, shown in Fig. 7.15.

If $a=6$, then $a+b+1=b+7$; if $a=8$, then $a+b+1=b+9$; so both of these cases yield a duplication. If $a=10$, there are cycles of length $e+11$ and $b+11$, so b and e cannot differ by 4; the only possibility is $\{b, e\} = \{1, 7\}$. Then

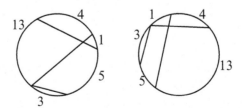

Fig. 7.15 Uniquely bipancyclic graphs of type ABC

Fig. 7.16 Chord type ACC

$b + e + 2 = 10$, and so does either $e + 9$ or $b + 9$. If $a = 10$, then $\{b, e\} = \{1, 7\}$. Then $b + e + 2 = 8$, and so does either $e + 7$ or $b + 7$. So there are no further uniquely bipancyclic graphs of type ABC.

Type ACC
A uniquely bipancyclic graph will contain 14 cycles, $n = 30$ and the chords total 238 edges. The cycles with 0 and 1 chord, the two XY cycles, the two YZ cycles, and the two XYZ cycles total 230 edges; the XZ cycle has $2 + c + d$, so $2 + c + d = 8$, $c + d = 6$. Without loss of generality, we can assume $c \le d$, so $c \le 3 \le d$. Each arc length is at least 1, so the only possible cycles of length 4 are the X cycle of length $(1 + a + b)$, the Y cycle of length $(1 + a + c + e)$, and the Z cycle of length $(1 + e + f)$; all others include at least five pieces (chords and arcs) or the arc of length d and at least two other pieces (see Fig. 7.16).

If $(1 + a + c + e) = 4$, then $a = c = e = 1$ and $d = 5$. One X cycle has length $1 + c + d + e + f = 8 + f$ and one XY cycle has length $2 + a + d + f = 8 + f$—a duplication. So there is no uniquely bipancyclic example with $(1 + a + c + e) = 4$.

If $(1 + a + b) = 4$, then either $a = 1, b = 2$, and the YZ cycle of length $2 + b + d + e$ is the same length as the XYZ cycle of length $3 + a + d + e$; or else $a = 2, b = 1$, and the YZ cycle of length $2 + a + c + f$ is the same length as the XYZ cycle of length $3 + b + c + f$—a duplication in either case. So there is no uniquely bipancyclic example with $(1 + a + b) = 4$. If $(1 + e + f) = 4$ then either $e = 1, f = 2$, and the XY cycle of length $2 + a + d + f$ is the same length as the XYZ cycle of length $3 + a + d + e$; or else $e = 2, f = 1$, and the XY cycle of length $2 + b + c + e$ is the same length as the XYZ cycle of length $3 + b + c + f$—a duplication in either case. So there is no uniquely bipancyclic example with $(1 + e + f) = 4$.

So in no case is there a uniquely bipancyclic graph of chord type ACC.

Type BBBi
This is impossible because it would contain a cycle of length 3.

Type BBBii
In this case, there are 10 cycles and 22 vertices, so the sum of the numbers of edges in the cycles is $4 + 6 + \cdots + 22 = 130$. There is one cycle of length 22 (the Hamilton circuit), six whose edges add to $3v + 6 = 72$ (the one-chord cycles), and cycles of lengths $a + 2$, $b + 2$ and $a + b + 2$, so the total number of edges in the cycles is $2a + 2b + 6 + 22 + 72 = 2(a + b) + 100$. So $a + b = 15$. But $a + 2$ and $b + 2$ must be even (because the XY and YZ chords are of equal lengths), so a and b must

Fig. 7.17 Chord type BBBii

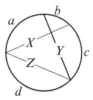

Fig. 7.18 Chord type BBC

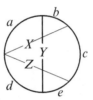

Fig. 7.19 Chord type BCC

be even—a contradiction. So there is no uniquely bicyclic graph of type BBBii (see Fig. 7.17).

Type BBC

Chord type BBC has 12 cycles, 26 vertices, and a total of 180 edges among the chords. The Hamilton cycle has 26 edges, the six one-chord cycles have 84 edges in total, and the two XY cycles total 30 edges. The XZ, YZ, and XYZ cycles have $c + 2$, $a + 2$ and $b + 3$ edges, respectively, so $a + b + c + 7 - 180 - 26 - 84 - 30 = 40$. So $a + b + c = 33$, which is impossible as there are only 26 vertices. Therefore chord pattern BBC is impossible (see Fig. 7.18).

Type BCC

A uniquely bipancyclic graph will contain 13 cycles, $n = 28$ and the chords total 208 edges. The cycles with 0 and 1 chord and the XY and YZ cycles total 182 edges; the XZ cycle has $2 + c$ and the XYZ cycle has $3 + b + e$. So $5 + b + c + e = 208 - 182 = 26$, $b + c + e = 21$, and $a + d = 7$. Without loss of generality we can assume $a < d$, so $a \leq 3$ (see Fig. 7.19).

If we use the substitutions $(7 - a)$ for d and $(21 - b - c)$ for e, the lengths of the non-Hamilton cycles are

$$\begin{array}{llll} a+b+1 & 29-a-b & a+b+c+1 & 29-a-b-c \\ 8 & 22 & 9+b-a & 23+a-b \\ 23+a-b-c & 9-a+b+c & 2+c & 24-c \end{array}$$

So $a + b$ is odd and c is even. Clearly c cannot equal 2 (there would be two cycles of length 22), whence $b + c \leq 17$; $a + b + c$ cannot equal 21 (two 22 cycles); and $a - b$ cannot equal ± 1 (two 8s or two 22s).

How can we achieve a cycle of length 4? The only possibilities are: $a + b = 3$, which would imply $a - b = \pm 1$; $a + b + c = 25$ (impossible as $b + c \leq 17$); $a - b = 5$ (impossible as $a \leq 3$); and $b + c = a + 19$ (again, impossible because $b + c \leq 17$). Therefore chord pattern BCC is impossible.

Type CCC

In a type CCC graph, there are six cycles with one chord, totalling $3(n + 2)$ edges, six with two chords, totalling $3(n + 4)$, two with three chords, contributing $n + 6$ edges, and the Hamilton cycle. So the cycle lengths add to

$$n + 3(n + 2) + 3(n + 4) + n + 6 = 8n + 24 = 280,$$

as $n = 32$. But the sum must be 270. So no solution is possible.

A bipancyclic graph with four chords will contain at least 15 cycles—two containing each chord, at least one for each pair of chords, plus the Hamilton cycle. So a uniquely bipancyclic graph with four chords will contain at least 32 vertices. Therefore we have found all uniquely bipancyclic graphs on 30 or fewer vertices.

There are uniquely bipancyclic graphs of orders, 4, 8, 14, and 26—examples of all of them are shown in Fig. 7.20, on the next page—and no other orders smaller than 32.

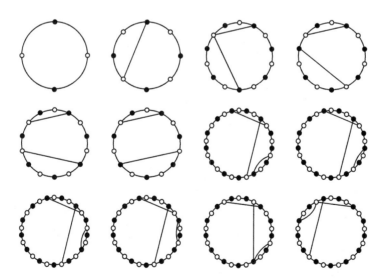

Fig. 7.20 The uniquely bipancyclic graphs of order less than 32

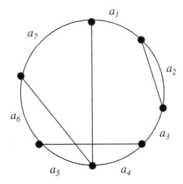

Fig. 7.21 Layout for the six non-isomorphic UBPC graphs of order 44

7.5 More Chords: Computer Searches

In this section by computer search we classify all UBPC graphs with four and five chords. Figure 7.21 displays a layout for possible UBPC graphs with four chords and two pairs of crossing chords. Note that there are precisely 21 different cycles in this layout. The order of a UBPC graph with 21 cycles is 44. The length of an arc on the Hamilton cycle is given by a_i, $1 \le i \le 7$. Note that if a_4, a_5, or a_6 is zero, then we will not have two pairs of crossing chords. So a_4, a_5, and a_6 are non-zero. If $a_7 = 0$, then we will have three chords. All UBPC graphs with three chords have been classified in Sect. 7.4. So $a_7 \ne 0$. In addition, we must have $a_2 \ge 3$ in order to avoid a repeated edge or a cycle of length 3. The lengths of the 21 cycles of this layout are given by:

$$\ell_1 = \sum_{i=1}^{7} a_i = 44;$$

$$\ell_2 = a_2 + 1;$$

$$\ell_3 = a_1 + 1 + a_3 + a_4 + a_5 + a_6 + a_7;$$

$$\ell_4 = a_4 + a_5 + 1;$$

$$\ell_5 = a_1 + a_2 + a_3 + 1 + a_6 + a_7;$$

$$\ell_6 = a_1 + a_2 + a_3 + a_4 + 1;$$

$$\ell_7 = a_5 + a_6 + a_7 + 1;$$

$$\ell_8 = a_1 + a_2 + a_3 + a_4 + 1 + a_7;$$

$$\ell_9 = a_5 + a_6 + 1;$$

$$\ell_{10} = a_1 + 1 + a_3 + a_4 + 1;$$

$$\ell_{11} = a_1 + 1 + a_3 + a_4 + 1 + a_7;$$

$$\ell_{12} = a_1 + 1 + a_3 + 1 + a_6 + a_7;$$

$$\ell_{13} = a_7 + 1 + 1;$$

$$\ell_{14} = a_4 + 1 + a_6 + 1;$$

$$\ell_{15} = a_5 + 1 + a_7 + a_1 + a_2 + a_3 + 1;$$

$$\ell_{16} = a_4 + 1 + a_7 + a_6 + 1;$$

$$\ell_{17} = a_1 + a_2 + a_3 + 1 + a_5 + 1;$$

$$\ell_{18} = a_1 + a_2 + a_3 + 1 + a_6 + 1 + 1;$$

$$\ell_{19} = a_1 + 1 + a_3 + 1 + a_5 + 1;$$

$$\ell_{20} = a_1 + 1 + a_3 + 1 + a_5 + 1 + a_7;$$

$$\ell_{21} = a_1 + 1 + a_3 + 1 + a_6 + 1 + 1.$$

A simple computer program shows that $\{\ell_i \mid 1 \leq i \leq 21\} = \{4, 6, 8, \ldots, 44\}$ if and only if $a_2 = 3$, $a_4 = 1$, $a_5 = 4$, $a_6 = 13$, $a_7 = 18$, and $a_1 + a_3 = 5$, where $0 \leq a_1 \leq 5$. Hence, this layout produces six non-isomorphic UBPC graphs of order 44. These are the graphs displayed in Fig. 7.21. They satisfy $a_1 + a_3 = 5$, $0 \leq a_1 \leq 5$, $a_2 = 3$, $a_4 = 1$, $a_5 = 4$, $a_6 = 13$, $a_7 = 18$.

In [21] the authors find 17 layouts with four chords and 75 layouts with 5 chords by hand for possible UBPC graphs. Based on the number of pairs of chords that intersect, the 17 layouts can be split into 7 groups and the 75 layouts into 11 groups. For each group the authors draw all possible layouts of chords that give the desired number of intersections. For each layout, they label each arc with a variable, where an arc is a path on the Hamilton cycle between adjacent chord endpoints (see Fig. 7.21). Then they use these variables to describe the length of each cycle with an equation. The cycles in a layout are found by computer search. The number of cycles indicates the order of that layout. Once they have these equations, they run a program that checks whether any combination of variable values results in a UBPC graph.

Note that a layout can simply be obtained from another layout by fixing some arc lengths at zero. Fixing an arc length at zero may reduce the number of cycles (and hence reduces the order) of a layout. Appendices A and B in [21] do not contain layouts with zero arc lengths. Any layout with zero arc lengths can be obtained from a layout in the appendices by fixing some arc lengths at zero. In addition, in a given layout the arc lengths are at zero only if this does not reduce the number of chords or the number of intersections.

We are now ready to state the main results obtained in [21].

Theorem 27. *1. There is no uniquely bipancyclic graph of order 2n, where $32 \leq 2n \leq 56$ and $2n \neq 44$.*
2. There are precisely six non-isomorphic uniquely bipancyclic graphs of order 44.

Chapter 8
Minimal Bipancyclicity

8.1 Introduction

Just as with pancyclic graphs, for every positive integer $n \geq 2$ there will be an integer such that any bipancyclic graph with $2n$ vertices must have at least that number of edges; in the case of bipancyclic graphs, we shall represent that integer as $m^*(2n)$ rather than $m(2n)$. A bipancyclic graph with $2n$ vertices and $m^*(2n)$ edges is called *minimal*. Recall that if a graph G has $e(G)$ edges and $v(G)$ vertices then the difference $e(G) - v(G)$ is called the *excess* of G, so $m^*(2n)$ is the value such that $m^*(2n) - 2n$ is the minimum excess for a bipancyclic graph on $2n$ vertices.

Again, we would conjecture that one cannot decrease $m^*(2n)$ by increasing v; in other words,

Conjecture 8.1. $m^*(2n) \geq m^*(2n - 2)$.

Just as in Sect. 3.1, we have an easy construction for bipancyclic graphs. Take a $2n$-cycle $(a_1, a_2, \ldots a_{2n}, a_1)$ and add edges from a_1 to a_i for $i = 4, 6, \ldots, n$ if n is even, and up to $n - 1$ if n is odd. The resulting graph is bipancyclic, but the excess is quite large for large n.

There has not been as much work on minimal bipancyclic graphs as on minimal pancyclic graphs. In the rest of this chapter we shall give some results on small orders. For convenience we shall denote the vertices of a bipartite graph of order $2n$ by $a_1, a_2, \ldots, a_{2n}$, where $(a_1, a_2, \ldots, a_{2n}, a_1)$ is the Hamilton cycle; the two components will be called V_1 and V_2, where V_1 consists of the vertices with odd subscripts and V_2 contains the even vertices.

© The Author(s) 2016
J.C. George et al., *Pancyclic and Bipancyclic Graphs*, SpringerBriefs
in Mathematics, DOI 10.1007/978-3-319-31951-3_8

Fig. 8.1 Minimal bipancyclic graphs up to order 8

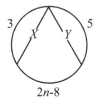

Fig. 8.2 Minimal bipancyclic graphs of orders 10–14

8.2   Minimal Bipancyclic Graphs with Excess Less than 2

It is obvious that the only bipancyclic graph on four vertices is the 4-cycle.

Suppose a bipancyclic graph has six vertices. Then it must contain cycles of length 4 and 6. The graph will contain a Hamilton cycle, so it will be a 6-cycle with additional edges. It is easy to add a 4-cycle by adding one edge, say from a_1 to a_4. The graph will in fact contain two 4-cycles.

As we saw in Sect. 7.2, there is a uniquely bipancyclic graph on 8 vertices. This graph and the examples on 4 and 6 vertices are shown in Fig. 8.1.

A graph consisting of a Hamilton cycle and one chord can contain at most three cycles. A bipancyclic graph on 10 or more vertices must contain at least four cycles, so it must contain at least two chords.

8.3   Excess 2

A graph with two chords will contain six or seven cycles—the Hamilton cycle, four containing exactly one chord (two for each), and one or two containing both chords, depending on whether or not they cross. So we need only consider cases $2n = 10$, 12, 14, and 16. A bipancyclic graph of order 16 with only two chords would necessarily be uniquely bipancyclic, and we saw in Sect. 7.3 that this is impossible.

Figure 8.2 shows a graph on $2n$ vertices with two chords, X and Y; the circle represents the Hamilton cycle, and the numbers on the cycle show the number of edges in the segment.

The Hamilton cycle is a cycle of length $2n$, the cycles containing X but not Y (the X-cycles) have lengths 4 and $2n - 2$, the Y-cycles have lengths 6 and $2n - 4$, and the XY-cycle is of length $2n - 6$. So the graph is bipancyclic for $2n = 10, 12$, or 14.

8.4   Excess 3

Suppose a Hamiltonian graph contains three chords. It will contain the Hamilton cycle, six one-chord cycles, three to six two-chord cycles, and one or two three-chord cycles, a total between 11 and 15 cycles. So a three-chord bipancyclic graph could have anywhere up to 32 vertices.

The possible arrangements of three chords were analyzed in Sect. 4.2.3 and were illustrated in Fig. 4.4. Figure 8.3 below shows a graph of type AAC on $2n$ vertices with three chords, X, Y, Z; the circle represents the Hamilton cycle, and the numbers on the cycle show the number of edges in the segment. The Hamilton cycle has length $2n$, the X-cycles have lengths $2n - 8$ and 10, the Y-cycles have length 6 and $2n - 4$, the Z-cycles have length 4 and $2n - 2$, the XY-cycles have length 14 and $2n - 10$, the XZ-cycle has length 8, the YZ-cycle has length $2n - 6$, and the XYZ-cycle has length 12. So the graph is bipancyclic for $10 \leq 2n \leq 26$.

A bipancyclic graph on 28 or more vertices must contain at least 13 cycles, so the only possible chord cases with three chords are BCC, ACC, or CCC (these patterns are defined in Fig. 7.5). Type BCC has 13 cycles, so a bipancyclic graph of that type on 28 vertices would be uniquely bipancyclic, which we saw in Sect. 7.4 to be impossible. The remaining cases are shown in Fig. 8.4, with the chords named and the lengths of arcs marked.

Type ACC has 14 cycles. In order to achieve 28 or more vertices, we can have at most one pair of cycles of the same length, but this cannot occur. The possible arc lengths to achieve a 4-cycle are $a = 2, b = 1$ or equivalent ($a = 1, b = 2$ or

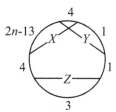

Fig. 8.3 Minimal bipancyclic graphs of orders 16–26

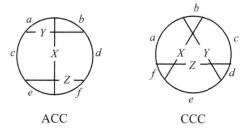

ACC   CCC

Fig. 8.4 Chord patterns ACC and CCC

$e = 1, f = 2$ or $e = 2, f = 1$), or $c = d = 1$. If $a = 2, b = 1$, there are an X-cycle and an XY-cycle of length $c + e + 3$, and an XZ-cycle and an XYZ-cycle of length $c + f + 4$; if $c = d = 1$, the XY-cycles are of lengths $b + e + 3$ and $a + f + 3$, and the XZ-cycles are of the same two lengths. In either case we have at most 12 distinct cycle lengths. So any bipancyclic graph on more than 26 vertices must have an excess of at least 4.

8.5   Excess 4

A graph with four chords could have as many as 31 cycles, so it is conceivable, though unlikely, that one might find a 4-chord bipancyclic graph with as many as 64 vertices. Very little work has been done on this topic. We know of a uniquely bipancyclic graph on 44 vertices, so we shall look at cases up to 44.

Figure 8.5 shows a pattern of graph with four chords and $2n$ vertices, where $2n$ must be at least 28. The numbers on the arcs are the number of edges. We list the lengths of the cycles according to the chords they contain:

No chords, $2n$;
A only, $10, 2n - 8$;   B only, $18, 2n - 16$;
C only, $4, 2n - 2$;   D only, $6, 2n - 4$;
A and B, $2n - 24$;   A and C, 8;
A and D, $14, 2n - 10$; B and C, $2n - 18$;
B and D, $16, 2n - 12$; C and D, $2n - 6$;
A, B, C, no cycle;   A, B, D, $24, 2n - 18$;
A, C, D, 12;   B, C, D, $2n - 14$;
A, B, C, D, 22.

So we always have cycles of all even lengths from 4 to 18, 22, and 24, $2n - 24$ and all orders from $2n - 20$ to $2n$. So the graph is bipancyclic provided $2n \leq 44$. (When $2n = 44$, the graph is the UBPC graph presented in Sect. 7.5.)

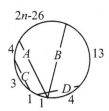

Fig. 8.5 A model for 28–44 vertices

We have

Theorem 28. *The minimum excess* $m^*(2n)$ *of a bipancyclic graph on* $2n$ *vertices satisfies* $m^*(4) = 0$, $m^*(6) = m^*(8) = 1$, $m^*(2n) = 2$ *for* $10 \leq 2n \leq 16$, $m^*(2n) = 3$ *for* $18 \leq 2n \leq 26$, $m^*(2n) = 4$ *for* $28 \leq 2n \leq 44$, *and* $m^*(2n) \geq 4$ *for* $2n \geq 46$.

8.6 More General Bounds for Bipancyclics

In this section we modify the construction given in [32] to obtain general bounds for minimal bipancyclic graphs (see also Sect. 4.5). For convenience, in this section we list the vertices as integers; for example, a graph of order $2n$ will have vertices $1, 2, \ldots, 2n - 1, 2n$.

Suppose n is an integer satisfying $2n \geq 4$ and write H_{2n} for the cycle $(1, 2, 3, \ldots, 2n, 1)$. Define $x_k = 2^k + k - 2$, for $k \geq 1$, and define the chord A_k in the cycle H_{2n} by $A_k = (x_k, x_{k+1})$ provided $2n \geq 2^{k+1} + k$. Note that the chord A_k has deficiency 2^k. We also define A_k^* to be the chord $(x_k, 1)$ if k is even and $(x_k, 2n)$ if k is odd. The deficiency of A_k^* is $2n - 2^k - k + 2$ when k is even and $2n - 2^k - k + 1$ when k is odd.

If $2n \geq 2^k + k - 2$, the chords $A_1, A_2, \ldots, A_{k-1}$ will not overlap. Add all those chords to H_{2n}. The resulting graph has no odd cycles, hence it is a bipartite graph by Theorem 2. This graph has cycles of all lengths $2n - d$, where $2 \leq d \leq 2^k - 2$ and ranges through the sums of deficiencies of the chords, as well as $2n$ and the orders $\{2^i + 2 \mid 1 \leq i \leq k - 1\}$.

We now add A_k^* to the graph. The resulting graph is still bipartite. When k is even, this introduces cycles of lengths $2n - 2^k - k + 4$ and $2^k + k - 2$. Provided $k \geq 2$, the cycle of length $2^k + k - 2$ encloses each of $A_1, A_2, \ldots, A_{k-1}$, hence we also get cycles of lengths $2^k + k - 2 - d$, where $2 \leq d \leq 2^k - 2$ and d ranges through the sums of deficiencies of the chords $A_1, A_2, \ldots, A_{k-1}$. In other words, we get all lengths from k to $2^k + k - 2$.

Similarly, when k is odd, this introduces cycles of lengths $2n - 2^k - k + 3$, $2^k + k - 1$, and $2^k + k - 1 - d$, where $2 \leq d \leq 2^k - 2$ and d ranges through the sums of deficiencies of the chords $A_1, A_2, \ldots, A_{k-1}$. In other words, we get all lengths from $k + 1$ to $2^k + k - 1$.

For positive even k define

$$G_k = H_{2n} \cup A_1 \cup A_2 \cup \cdots \cup A_{k-1} \cup A_k^*$$

for all orders $2n$ from $2^k + k + 2$ to $2^{k+1} + k$, and

$$G_k = H_{2n} \cup A_1 \cup A_2 \cup \cdots \cup A_{k-1} \cup A_k$$

when $2n = 2^{k+1} + k + 2$.

Similarly, for positive odd k define,

$$G_k = H_{2n} \cup A_1 \cup A_2 \cup \cdots \cup A_{k-1} \cup A_k^*$$

for all orders $2n$ from $2^k + k + 1$ to $2^{k+1} + k + 1$, and

$$G_k = H_{2n} \cup A_1 \cup A_2 \cup \cdots \cup A_{k-1} \cup A_k$$

when $2n = 2^{k+1} + k + 3$.

Start with a Hamilton cycle $H_{2n} = (1, 2, \ldots, 2n, 1)$, $2n \geq 4$ and even, which is itself a bipancyclic graph when $2n = 4$. Add the chord $A_1 = (1, 4)$. The resulting graph is bipancyclic when $2n = 6, 8$. Now add the chord $A_2 = (4, 9)$. This provides a bipancyclic graph for $2n = 10, 12, 14$. The next case is $2n = 16$ and we add the chord $A_3^* = (9, 16)$ to obtain a bipancyclic graph. When $2n = 18, 20$, or 22 the graph $H_{2n} \cup A_1 \cup A_2 \cup A_3$, where $A_3 = (9, 18)$, is bipancyclic. Next we add $A_4^* = (2n, 1)$. The resulting graph has the four chords A_1, A_2, A_3, and A_4^* and is bipancyclic graph for $2n = 24, 26, \ldots, 36$. If $2n = 38$, this graph is not bipancyclic because it misses a cycle of length 20. The graph $H_{38} \cup A_1 \cup A_2 \cup A_3 \cup A_4$, where $A_4 = (18, 35)$, is bipancyclic but the graph $H_{40} \cup A_1 \cup A_2 \cup A_3 \cup A_4$ is missing a cycle of length 8. We have bipancyclic graphs with the following parameters:

Vertices	Graph	Edges	Excess
$2n = 4$	H_4	4	0
$6 \leq 2n \leq 8$	G_1	$2n + 1$	1
$10 \leq 2n \leq 14$	G_2	$2n + 2$	2
$16 \leq 2n \leq 22$	G_3	$2n + 3$	3
$24 \leq 2n \leq 38$	G_4	$2n + 4$	4

When $2n \geq 40$ we add both A_4 and A_4^*. So our graph is $H_{2n} \cup A_1 \cup A_2 \cup A_3 \cup A_4 \cup A_4^*$. Chords A_1, A_2, A_3, A_4 provide cycles of lengths 4, 6, 10, 18, and $2n - d : 2 \leq d \leq 2^5 - 2 = 30$ and d even. H_{2n} provides length $2n$; and A_1, A_2, A_3, A_4^* provide even orders from 4 to $4 + (2^4 - 2) = 18$. So the resulting graph is pancyclic provided $2n - 30 \leq 20$. So we have a pancyclic graph of excess 5 when $40 \leq 2n \leq 50$. We also noticed that when $2n = 52$, the graph $H_{2n} \cup A_1 \cup A_2 \cup A_3 \cup A_4 \cup A_4^*$ also has the cycle $(1, 4, 9, 18, 35, 36, 37, \ldots, 52, 1)$, which is of length 22, so it is bipancyclic.

The graph $G_j = A_1 \cup A_2 \cup \cdots \cup A_j$ has cycles of lengths $2^i + 2$, for $2 \leq i \leq j$, and cycles of all lengths from $2n - (2^{j+1} - 2)$ upwards. The addition of A_4^* to G_j provides cycles of all even lengths from 4 up to 18. So, for $4 \leq j \leq 18$, $G_j \cup A_4^*$ is a bipancyclic graph with an excess of $j + 1$ provided $2n - (2^{j+1} - 2) \leq 20$; that is, $2n \leq 2^{j+1} + 18$, and provided none of $A_1, A_2, \ldots, A_j$ overlap. So we have a bipancyclic graph of excess $j + 1$ when $2^j + j \leq 2n \leq 2^{j+1} + 18$.

Note that the "+18" addition to the upper limit of $2n$ means that a few of the smallest values for which we just established an excess of $j + 1$ will in fact be covered by the construction for excess j, so we actually have

Theorem 29. *When* $4 \le j \le 20$, *there is a bipancyclic graph on* $2n$ *vertices with* $2n + j + 1$ *edges whenever* $2^j + 20 \le 2n \le 2^{j+1} + 18$.

Of course, this theorem is clearly not best possible; in fact, it is not quite as good as Theorem 28. But we now have a bound for all orders up to $2n = 2^{21} + 18 = 2,097,170$. Here are some of the smaller cases: we have found bipancyclic graphs with the following parameters:

Vertices	Edges	Excess
$36 \le 2n \le 50$	$2n + 5$	5
$52 \le 2n \le 82$	$2n + 6$	6
$84 \le 2n \le 146$	$2n + 7$	7
$148 \le 2n \le 274$	$2n + 8$	8
$276 \le 2n \le 530$	$2n + 9$	9
$532 \le 2n \le 1042$	$2n + 10$	10

When $j = 22$, there will be no cycle of length 20. However, the addition of the chord $A_5^* = (35, 2n)$ provides cycles of all lengths from 6 to 36. Note that when the chord A_j^* is added, the chords $A_1, A_2, \ldots, A_{j-1}, A_j^*$ together form cycles of all length from j to $2^j + j - 2$ if j is even and from $j + 1$ to $2^j + j - 1$ if j is odd. In general, to avoid missing a cycle of length $2^h + h$, where h is even, chord A_{h+2}^* is needed when $2n \ge 2^j + j + 2$, where $j = 2^h + h$. This means that when $2^{(2^h+h)} + 2^h + h + 2 \le 2n \le 2^{(2^h+h+1)} + 2^h + h$ we add $A_1, A_2, \ldots, A_{(2^h+h)}$ and $A_4^*, A_5^*, \ldots, A_{(h+2)}^*$ to H_{2n} to achieve a bipancyclic graph.

So we have

Theorem 30. *When*

$$2^{(2^h+h)} + 2^h + h + 2 \le 2n \le 2^{(2^h+h+1)} + 2^h + h,$$

there is a bipancyclic graph on $2n$ *vertices with* $2n + 2^h + 2h - 1$ *edges. So the excess for a minimal bipancyclic graph with* $2n$ *vertices as stated is at most* $2^h + 2h - 1$.

8.7 Bipancyclic Graph Products

In Sect. 3.3 we discussed when certain products of graphs exhibit pancyclicity. As we said, the *cartesian product*, denoted $G \times H$, has an edge between $g_1 h_1$ and $g_2 h_2$ when either ($g_1 = g_2$ and $h_1 \sim h_2$) or ($h_1 = h_2$ and $g_1 \sim g_2$). Of particular interest is the construction $G \times K_2$, called a *prism*. In that section, we cited the following result from [14], as Theorem 15:

Theorem 31. *If* G *is a 3-connected cubic graph on* $2n$ *vertices, then* $G \times K_2$ *has cycles of every even length from 4 up to 4n; and if in addition* G *contains a 3-cycle, then* $G \times K_2$ *is pancyclic.*

Since the product $G \times H$ is bipartite if G and H are bipartite, we have an immediate corollary:

Corollary 31.1. *If G is a bipartite 3-connected cubic graph, then $G \times K_2$ is bipancyclic.*

As we saw in Sect. 1.3, the *tensor product*, also called the Kronecker product or conjunction, is the product $G \otimes H$ with an edge between $g_1 h_1$ and $g_2 h_2$ when $(g_1 \sim g_2$ and $h_1 \sim h_2)$. The product possesses two properties of interest here. First, if either G or H are bipartite, then $G \otimes H$ is also bipartite. Second, if both are bipartite, then the product is disconnected. (If G and H are connected and bipartite, then $G \otimes H$ has exactly two components.) The following theorem is a result of Jha [20]:

Theorem 32. *The product $C_m \otimes C_{4j}$ has a bipancyclic ordering if m is odd, and each component has a bipancyclic ordering if m is even.*

This statement is slightly stronger than the usual statement that the graph is bipancyclic; it asserts that there exists an ordering of the vertices $v_0, v_1, \ldots$ with the property that $(v_0, v_1, \ldots, 2\ell - 1)$ is a cycle of length 2ℓ for each ℓ up to the order of the graph.

Regarding prisms, it follows from Corollary 31.1 that if G is bipartite, cubic, and 3-connected, then the prism $G \times K_2$ is bipancyclic.

References

1. D. Amar, E. Flandrin, I. Fournier, A. Germa, Pancyclism in Hamiltonian graphs. Discrete Math. **89**, 111–131 (1991)
2. D. Bauer, E. Schmeichel, Hamiltonian degree conditions which imply a graph is pancyclic. J. Comb. Theory (B) **48**, 111–116 (1990)
3. J.A. Bondy, Pancyclic graphs I. J. Comb. Theory (B) **11**, 80–84 (1971)
4. J.A. Bondy, Pancyclic graphs: recent results. Colloq. Math. Soc. János Bolyai, 181–187 (1973)
5. J.A. Bondy, Longest paths and cycles in graphs of high degree. Research Report CORR 80-16, University of Waterloo, Waterloo, Ontario (1980)
6. R.L. Brooks, On coloring the nodes of a graph. Proc. Camb. Philos. Soc. **37**, 194–197 (1941)
7. W.K. Chen, On vector spaces associated with a graph. SIAM J. Appl. Math. **20**, 385–389 (1971)
8. V. Chvátal, On Hamilton's ideals. J. Comb. Theory (B) **12**, 163–168 (1972)
9. G.A. Dirac, Some theorems on abstract graphs. Proc. Lond. Math. Soc. **2**, 69–81 (1952)
10. R.C. Entringer, E.F. Schmeichel, Edge conditions and cycle structure in bipancyclic graphs . Ars. Comb. **26**, 229–232 (1988)
11. G.-H. Fan, New sufficient conditions for cycles in graphs. J. Comb. Theory (B) **37**, 221–227 (1984)
12. I. Fournier, P. Fraisse, One a conjecture of Bondy. J. Comb. Theory (B) **39**, 17–26 (1985)
13. J.C. George, A. Marr, W.D. Wallis, Minimal pancyclic graphs. J. Comb. Math. Combin. Comput. **86**, 125–133 (2013)
14. W. Goddard, M.A. Henning, Note: pancyclicity of the prism. Discrete Math. **234**, 139–142 (2001)
15. R. Gould, Graphs and vector spaces. J. Math. Phys. **37**, 193–214 (1958)
16. S. Griffin, Minimal pancyclicity (to appear)
17. R. Häggkvist, Odd cycles of specified length in nonbipartite graphs. Ann. Discrete Math. **62**, 89–99 (1982)
18. W. Imrich, S. Klavzar, *Product Graphs, Structure and Recognition* (Wiley, New York, 2000)
19. S. Janson, T. Łuczak, A. Ruciński, *Random Graphs* (Wiley, New York, 2000)
20. P.K. Jha, Kronicker products of paths and cycles: decomposition, factorization, and bi-pancyclicity. Discrete Math. **182**, 153–167 (1998)
21. A. Khodkar, A. Peterson, C. Wahl, Z. Walsh, Uniquely bipancyclic graphs on more than 30 vertices. J. Comb. Math. Combin. Comput. (to appear)
22. C. Lai, Graphs without repeated cycle lengths. Aust. J. Comb. **27**, 101–105 (2003)
23. K. Markström, A note on uniquely pancyclic graphs. Aust. J. Comb. **44**, 105–110 (2009)

© The Author(s) 2016
J.C. George et al., *Pancyclic and Bipancyclic Graphs*, SpringerBriefs in Mathematics, DOI 10.1007/978-3-319-31951-3

24. J. Mitchem, E.F. Schmeichel, Pancyclic and bipancyclic graphs – a survey, in *Proceedings of the First Colorado Symposium on Graph Theory*, ed. by F. Harary, J.S. Maybee (Wiley, New York, 1985), pp. 271–278.
25. O. Ore, Note on Hamilton circuits. Am. Math. Month. **67**, 55 (1960)
26. N.C.K. Phillips, W.D. Wallis, Uniquely bipancyclic graphs on thirty-two vertices. J. Discrete Math. Sci. Crypt. (to appear)
27. S. Ramachandran, R. Parvathy, Pancyclicity and extendability in strong products. J. Graph Theory **22**, 75–82 (1996)
28. E.F. Schmeichel, S.L. Hakimi, Pancyclic graphs and a conjecture of Bondy and Chvatal. J. Comb. Theory (B) **17**, 22–34 (1974)
29. E.F. Schmeichel, J. Mitchem, Bipartite graphs with cycles of all even lengths. J. Graph Theory **6**, 429–439 (1982)
30. Y. Shi, Some theorems of uniquely pancyclic graphs. Discrete Math. **59**, 167–180 (1986)
31. Y. Shi, The number of cycles in a Hamilton graph. Discrete Math. **133**, 249–257 (1994)
32. M.R. Sridharan, On an extremal problem concerning pancyclic graphs. J. Math. Phys. Sci. **12**, 297–306 (1978)
33. K. Thulasiraman, M.N.S. Swamy, *Graphs: Theory and Algorithms* (Wiley, New York, 1992)
34. W.D. Wallis, *A Beginner's Guide to Graph Theory*, 2nd edn. (Birkhäuser, Boston, MA, 2007)
35. W.D. Wallis, Uniquely bipancyclic graphs. J. Comb. Math. Comb. Comput. (to appear)
36. D.B. West, *An Introduction to Graph Theory*, 2nd edn. (Prentice-Hall, Englewood Cliffs, NJ, 2001)
37. H. Whitney, On the abstract properties of linear dependence. Am. J. Math. **57**, 509–533 (1935)
38. C.T. Zamfirescu, (2)-Pancyclic graphs. Discrete Appl. Math. **161**, 1128–1136 (2013)

Printed in the United States
By Bookmasters